Identity Management for Internet of Things

RIVER PUBLISHERS SERIES IN COMMUNICATIONS
Volume 39

Consulting Series Editors

MARINA RUGGIERI
University of Roma "Tor Vergata"
Italy

HOMAYOUN NIKOOKAR
Delft University of Technology
The Netherlands

ABBAS JAMALIPOUR
The University of Sydney, Australia

This series focuses on communications science and technology. This includes the theory and use of systems involving all terminals, computers, and information processors; wired and wireless networks; and network layouts, procontentsols, architectures, and implementations.

Furthermore, developments toward newmarket demands in systems, products, and technologies such as personal communications services, multimedia systems, enterprise networks, and optical communications systems.

- Wireless Communications
- Networks
- Security
- Antennas & Propagation
- Microwaves
- Software Defined Radio

For a list of other books in this series, visit www.riverpublishers.com
http://riverpublishers.com/river publisher/series.php?msg=Communications

Identity Management for Internet of Things

Parikshit N. Mahalle
Poonam N. Railkar

*Department of Computer Engineering,
Sinhgad Technical Educations Society's
Smt. Kashibai Navale College of Engineering,
Savitribai Phule Pune University, Pune, India*

LONDON AND NEW YORK

Published 2015 by River Publishers
River Publishers
Alsbjergvej 10, 9260 Gistrup, Denmark
www.riverpublishers.com

Distributed exclusively by Routledge
4 Park Square, Milton Park, Abingdon, Oxon OX14 4RN
605 Third Avenue, New York, NY 10158

First published in paperback 2024

Identity Management for Internet of Things / by Parikshit N. Mahalle, Poonam N. Railkar.

© 2015 River Publishers. All rights reserved. No part of this publication may be reproduced, stored in a retrieval systems, or transmitted in any form or by any means, mechanical, photocopying, recording or otherwise, without prior written permission of the publishers.

Routledge is an imprint of the Taylor & Francis Group, an informa business

Publisher's Note
The publisher has gone to great lengths to ensure the quality of this reprint but points out that some imperfections in the original copies may be apparent.

While every effort is made to provide dependable information, the publisher, authors, and editors cannot be held responsible for any errors or omissions.

ISBN: 978-87-93102-90-3 (hbk)
ISBN: 978-87-7004-489-9 (pbk)
ISBN: 978-1-003-33850-5 (ebk)

DOI: 10.1201/9781003338505

Contents

Preface ix

Acknowledgements xiii

List of Figures xv

List of Tables xvii

List of Acronyms xix

1 Internet of Things Overview 1
   1.1 Overview 1
       1.1.1 Internet of Things: Vision 2
       1.1.2 Emerging Trends 4
       1.1.3 Economic Significance 6
   1.2 Technical Building Blocks 8
       1.2.1 Internet of Things Layered Architecture 9
       1.2.2 RFID and Internet of Things 12
       1.2.3 IP for Things 16
   1.3 Issues and Challenges 18
       1.3.1 Design Issues 18
       1.3.2 Technological Challenges 20
       1.3.3 Security Challenges 21
   1.4 Applications 22
       1.4.1 Manufacturing, Logistic and Relay 22
       1.4.2 Energy and Utilities 23
       1.4.3 Intelligent Transport 23
       1.4.4 Environmental Monitoring 23
       1.4.5 Home Management 23
       1.4.6 eHealth 24

vi   Contents

	1.5	Conclusions	24
		References	25
2	**Elements of Internet of Things Security**		**29**
	2.1	Introduction	29
		2.1.1 Vulnerabilities of IoT	29
		2.1.2 Security Requirements	33
		2.1.3 Challenges for Secure Internet of Things	34
	2.2	Threat Modeling	37
		2.2.1 Threat Analysis	37
		2.2.2 Use Cases and Misuse Cases	38
		2.2.3 Activity Modeling and Threats	40
		2.2.4 IoT Security Tomography	43
	2.3	Key Elements	46
		2.3.1 Identity Establishment	47
		2.3.2 Access Control	48
		2.3.3 Data and Message Security	48
		2.3.4 Non-repudiation and Availability	49
		2.3.5 Security Model for IoT	49
	2.4	Conclusions	51
		References	51
3	**Identity Management Models**		**55**
	3.1	Introduction	55
		3.1.1 Identity Management	55
		3.1.2 Identity Portrayal	59
		3.1.3 Related Works	61
	3.2	Different Identity Management Models	64
		3.2.1 Local Identity	64
		3.2.2 Network Identity	66
		3.2.3 Federated Identity	67
		3.2.4 Global Web Identity	69
	3.3	Identity Management in Internet of Things	69
		3.3.1 User-centric Identity Management	70
		3.3.2 Device-centric Identity Management	71
		3.3.3 Hybrid Identity Management	71
	3.4	Conclusions	74
		References	75

4 Identity Management and Trust — 79
- 4.1 Introduction — 79
  - 4.1.1 Motivation — 81
  - 4.1.2 Trust Management Life Cycle — 82
  - 4.1.3 State of the Art — 83
- 4.2 Identity and Trust — 85
  - 4.2.1 Third Party Approach — 85
  - 4.2.2 Public Key Infrastructure — 86
  - 4.2.3 Attribute Certificates — 87
- 4.3 Web of Trust Models — 90
  - 4.3.1 Web Services Security — 92
  - 4.3.2 SAML Approach — 94
  - 4.3.3 Fuzzy Approach for Trust — 95
- 4.4 Conclusions — 97
- References — 98

5 Identity Establishment — 103
- 5.1 Introduction — 103
  - 5.1.1 Mutual Identity Establishment in IoT — 104
  - 5.1.2 IoT Use Case and Attacks Scenario — 104
  - 5.1.3 State of the Art — 107
- 5.2 Cryptosystem — 109
  - 5.2.1 Private Key Cryptography — 111
  - 5.2.2 Public Key Cryptography — 111
- 5.3 Mutual Identity Establishment Phases — 113
  - 5.3.1 Secret Key Generation — 113
  - 5.3.2 One Way Authentication — 114
  - 5.3.3 Mutual Authentication — 115
- 5.4 Comparative Discussion — 115
  - 5.4.1 Security Protocol Verification Tools — 116
  - 5.4.2 Security Analysis — 118
  - 5.4.3 Performance Metrics — 122
- 5.5 Conclusions — 124
- References — 124

6 Access Control — 129
- 6.1 Introduction — 129
  - 6.1.1 Motivation — 130
  - 6.1.2 Access Control in Internet of Things — 131

		6.1.3 Different Access Control Schemes	133
		6.1.4 State of the Art	134
	6.2	Capability-based Access Control	135
		6.2.1 Concept of Capability	136
		6.2.2 Identity-based Capability Structure	137
		6.2.3 Identity-driven Capability-based Access Control	138
	6.3	Implementation Considerations	139
		6.3.1 Functional Specifications	141
		6.3.2 Access Control Policies Modeling	142
		6.3.3 Mobility and Backward Compatibility	145
	6.4	Conclusions	147
		References	148
7	**Conclusions**		**153**
	7.1	Summary	153
	7.2	Identity Management Framework	156
	7.3	Future Outlook	158
		References	159
Index			**161**
About the Authors			**163**

Preface

Intelligence refers to the power to analyze things in their proper perspective, and knowledge refers to understanding what is spirit and what is matter.

(Bhagavad Gita 10.4)

This is intended to be simple, accessible book and primary reference which puts forwards best research roadmap, challenges and future outlook on Identity Management in Internet of Things. As Einstein said, "Everything should be made as simple as possible, but no simpler." Essentially, this point of view drives our development throughout the book. Applications of Internet of Things are growing tremendously in the domains of habitat, tele-health, industry monitoring, vehicular networks, home automation and agriculture. In all the domains, identity and access management is an important challenge to be addressed. This trend is strong motivation to address Identity Management issues because identity portrayal, secure identification of users, host and devices is considered the core element of Internet of Things security.

The main characteristics of this book are:

- It assumes that the reader's goal is to achieve a complete understanding of Identity Management issues, challenges and possible solutions in Internet of Things. It is not oriented towards any specific application and Identity Management problem is discussed regardless of the certain OSI layer.
- This book is motivating to use Internet of Things paradigm in new inventions for wide range of stakeholders like layman to educated users, villages to metros and national to global levels.
- This book contains numerous examples, case studies, technical descriptions, procedures, algorithms and protocols. These deliverables have been developed with the greatest of care and they will be useful to the readers in a broad range of applications.
- Another endeavor of this book is modeling of threats and attacks using use case approach in order to give an actual view of the Internet of Things.

This approach is useful to the reader for better understanding and further investigations.

Chapter 1 gives an understanding of the Internet of Things, its vision and economic significance. The role of radio frequency identification, its expected business relevance is also introduced in this chapter. Finally Chapter 1 concludes with different design issues, technological and security challenges as well as different applications and usage scenarios in stakeholder perspective.

Chapter 2 presents vulnerabilities in Internet of Things and discusses the requirements and challenges for handling successful security. In this chapter, threat modeling, threat analysis and use cases and misuse cases are also discussed in general. In the sequel, security tomography and security model is discussed in the last part of this chapter.

Chapter 3 describes concept of identity and the identity portrayal in Internet of Things as well as existing identifier schemes. Different Identity Management models and the requirements of Identity Management solutions are discussed in the last part of this chapter.

Chapter 4 explains the importance of trust for Identity Management and access control. A relationship between trust and access control is introduced and trust management life cycle is also discussed in this chapter. The main paradigms for identity trust, different web of trust models and emerging mechanisms for the exchange of security constructs are discussed in the last part of this chapter.

Chapter 5 initially presents motivation and challenges of authentication in Internet of Things. This chapter also presents attack modeling using use cases, and the threat analysis for these attacks. Finally chapter 5 concludes with an overview of cryptosystem and its applicability in Internet of Things, discussion on proposed mutual identity establishment scheme and its performance analysis.

Chapter 6 describes access control issues, motivation and challenges specific to Internet of Things using use cases. Different access control schemes and its evaluation are presented in the next part of this chapter. Capability-based authorization approach for management of access control and its functionality is explained at the end of this chapter.

Finally, Chapter 7 summarizes the book and concludes by proposing the future work, which can be researched and build based on the ideas and challenges presented. This chapter also presents Identity Management framework as an integration of different building blocks of Identity Management presented in this book.

The book is useful for Undergraduates, Postgraduates, Industry, Researchers and Research Scholars in ICT and we are sure that, this book will be well received by all stakeholders.

Parikshit N. Mahalle
Poonam N. Railkar

Acknowledgements

I would like to acknowledge and appreciate the support of my co-author Poonam N. Railkar for her major contribution towards improving the value of this book. I am very much thankful to Poonam for being proactive and instrumental in providing her thoughts and suggestions to address best practices in this book, wonderful coordination and generous response. I am proud to write a few words to present Poonam as the technical reviewer of this book. I would like to acknowledge and thank my Ph. D supervisor Associate Professor Neeli Rashmi Prasad immensely for her tireless and unconditional help, support, guidance and being a role model for me. I am highly indebted to Professor Ramjee Prasad and Dr. S. S. Inamdar for their encouragement and inspirational support. I am also thankful to my Master student Nancy for her help and support.

I would like to thank my mother for supporting my academic endeavours. I am thankful to her for being a backbone to me always. I am deeply indebted to my father Late Dr. Narendra Mahalle for his inspiration and love. He would have been very happy today.

Lastly but importantly, I wants to thank to my loving wife Dr. Namita. She has tirelessly supported me while writing up this book. I would also like to thank my daughter Yashita for making me forget all the pressure with her innocent smile. I would also like to thanks all those who directly and indirectly involved in building this book and research work.

Parikshit N. Mahalle

At first, I would like to thank whole heartedly Dr. Parikshit N. Mahalle for his generous support and constant encouragement to complete this Book. I have learnt so much technical things from him. I take this opportunity to express my deep sense of gratitude towards my esteemed Head of the department and co-author Dr. Parikshit N. Mahalle for giving me this splendid opportunity to write this book.

I would like to express lot of thanks to my loving husband Mr. Ninad for his generous support. His sacrifices allow me to use my spare time constructively and fruitfully.

I would like to express lot of thanks to my little and wonderful kids Anish and Advait for making me forget all the difficulties with their smiles and support. I would like to extend my thanks to my Mother-In-Law Mrs. Deepali V. Railkar for her help and love. I would like to thank my students Nancy and Pawan for their support and help. Last but not the least I would like to thank my Parents, whole family and Friends also for understanding and supportive in the process of creating this book.

Poonam N. Railkar

We could not have written this book without the encouragement of our well-wishers and family members. We would like to thank many people who encouraged and helped us in various ways throughout this book, namely our colleagues, friends and students.

We are indebted to Honourable founder president of STES, Prof. M. N. Navale, founder secretary of STES, Dr. Mrs. S. M. Navale, Vice President (HR), Mr. Rohit M. Navale, Vice President (Admin), Ms. Rachana M. Navale, our Principal Dr. A. V. Deshpande, Vice Principal Dr. K. R. Borole, Dr. K. N. Honwadkar for their constant encouragement, inexplicable support and faith on us. We are also very much thankful to all my department colleagues at SKNCOE for their continuous support, help and keeping us smiling all the time.

Last but not the least; our acknowledgements would remain incomplete if we do not thank the team of River Publications, Aalborg, Denmark who supported us throughout the development of this book. It has been pleasure to work with the River Publications team and we extend our special thanks to entire team involved in the publication of this book.

Finally, we would like to thank Mr. Rajeew Prasad, Elizabeth Jansen and Junko Nakajima of River Publications, Aalborg for their continuous support in completing this book.

Parikshit N. Mahalle
Poonam N. Railkar

List of Figures

Figure 1.1	The big picture	2
Figure 1.2	IoT applications and functional blocks	8
Figure 1.3	High level layered architecture of IoT	11
Figure 1.4	RFID tag-reader communication	13
Figure 1.5	EPCGlobal overall set of architecture framework components and layers	14
Figure 1.6	RFID system architecture	17
Figure 1.7	IoT design issues	19
Figure 2.1	The life cycle of a device in IoT	30
Figure 2.2	Vulnerabilities of IoT	31
Figure 2.3	Security architecture for IoT	35
Figure 2.4	IoT use case	41
Figure 2.5	IoT attacks scenario	43
Figure 2.6	IoT security tomography and layered attacker model	44
Figure 2.7	Security model for IoT	50
Figure 3.1	Things and identifiers in IoT	57
Figure 3.2	Identifier format for things	58
Figure 3.3	Identity portrayal	60
Figure 3.4	High level view of local identity management model	65
Figure 3.5	High level of network identity model	67
Figure 3.6	An example of federated IoT network	68
Figure 4.1	Trust management life cycle	83
Figure 4.2	PKI trust topologies	88
Figure 4.3	X.509 v2 AC	89
Figure 4.4	WoT as directed graph	91
Figure 4.5	Web service architecture	92
Figure 4.6	SAML components	96
Figure 5.1	General process of mutual identity establishment	105
Figure 5.2	Use case for denial of service attack	106
Figure 5.3	Use case for man-in-middle attack	107
Figure 5.4	Use case for replay attack	108

Figure 5.5	High level view of private key cryptography	111
Figure 5.6	High level view of of public key cryptography	112
Figure 5.7	ECCDH for establishing shared secret key	114
Figure 5.8	One way authentication protocol	116
Figure 5.9	Protocol for mutual authentication	117
Figure 5.10	Attack on identity/location privacy	121
Figure 6.1	Use case for gateway registration	131
Figure 6.2	Use case for RPS	132
Figure 6.3	ACL versus capability-based access control	137
Figure 6.4	Identity driven capability structure	138
Figure 6.5	High level functioning of CAC	139
Figure 6.6	CAC scheme for IoT	140
Figure 6.7	Use case for connection establishment	142
Figure 6.8	Use case for ICAP generation	143
Figure 6.9	Use case for sending ICAP	143
Figure 6.10	Use case for receiving ICAP	144
Figure 6.11	Wireless system evolution	146
Figure 7.1	IdM architecture	157
Figure 7.2	IdM framework	158

List of Tables

Table 2.1	Use case and misuse case for access control	38
Table 3.1	Limitations of different identification schemes	62
Table 3.2	Comparative summary of the state of the art for identity management	73
Table 4.1	Difference between AC and PKC	89
Table 5.1	State of the art evaluation summary	110
Table 5.2	Computational time for MIECAC	123
Table 6.1	Comparison of different access control models	134

List of Acronyms

AA	Attribute Authority
ABAC	Attribute-Based Access Control
AC	Attribute Certificate
ACL	Access Control List
ACM	Access Control Matrix
AES	Advanced Encryption Standard
AGW	Access Gateway
ASN	Abstract Syntax Notation
AVISPA	Automated Validation of Internet Security Protocols and Applications
BAC	Building, Automation and Control
BTNS	Better-Than-Nothing Security
CA	Certification Authority
CAC	Capability-based Access Control
CCHA	Context aware Clustering with Hierarchical Addressing
CID	Context Identity
CL	Capability List
Cl-Atse	Constraint-Logic-based Attack searcher
CRBAC	Context-aware Role-Based Access Control
CWAC	Context-Aware Access Control
DDoS	Distributed Denial of Service
DH	Diffie-Hellman
DHCP	Dynamic Host Configuration Protocol
DHT	Distributed Hash Table
DNS	Domain Name System
DOS	Denial of service
DTAP	Direct Transfer Application Part
ECC	Elliptic Curve Cryptography
ECCDH	Elliptical Curve Cryptography-Diffie Hellman
EPC	Electronic Product Code
EPCIS	EPC Information Service
EU	European Union
F&C	Filtering & Collection
FP6	Sixth Framework Programme
HLPSL	High Level Protocol Specification Language
HTTP	Hypertext Transfer Protocol
ICAP	Identity-based Capability
IdM	Identity Management

IETF	Internet Engineering Task Force	
IF	Intermediate Format	
IoT	Internet of Things	
ITU	International Telecommunications Union	
KDC	Key Distribution Center	
MAC	Message Authentication Code	
MIECAC	Mutual Identity Establishment and Capability-based Access Control	
NR	Non-repudiation	
OFMC	On the Fly Model Checker	
ONS	Object Name Service	
PCC	Policy and Charging Control	
PCRF	Policy and Charging Rule Function	
PGP	Pretty Good Policy	
PKC	Public Key Certificate	
PUKC	Public Key Cryptography	
PKI	Public Key Infrastructure	
PML	Physical Markup Language	
POP	Proof of Possession	
RBAC	Role-Based Access Control	
RFID	Radio Frequency Identification	
RPS	Remote Printing Service	
RSA	Rivest-Shamir-Adelman	
SATMC	SAT base Model Checker	
SGTIN	Serialized Global Trend Identifier	
SKC	Secret Key Cryptography	
SOA	Services Oriented Architecture	
SOAP	Simple Object Access Protocol	
SSO	Single Sign On	
STA	Symbolic Trace Analyzer	
TA4SP	Tree Automata Based on Automatic Approximation for the Analysis of Security Protocols	
TCGA	Threshold Cryptography-based Group Authentication	
TCP	Transmission Control Protocol	
URI	Universal Resource Identifier	
URL	Uniform Address Locator	
WoT	Web of Trust	
WS	Web Services	
WSDL	Web Services Description Language	
WSN	Wireless Sensor Network	
WWW	World Wide Web	
XML	Extensible Markup Language	

1
Internet of Things Overview

1.1 Overview

Numbers of changes in the application and communication technology have been witnessed since the evolution of Internet almost 30 years ago. The Internet has undergone severe changes since its first launch in the late 1960s as an outcome of the ARPANET with number of users about 20% of the world population. The Internet of past is transformed to service oriented ubiquitous infrastructure due to anything, anytime and anywhere operations. In such ambient environment not only user become ubiquitous but also devices and their context become transparent and ubiquitous. With the miniaturization of devices, increase of computational power, and reduction of energy consumption, this trend will continue in the near future.

The consumers of today's networked world are swamped with information coming from a myriad of applications and services present on their devices, communication infrastructures and on the Internet. In the near future, the information overload will be magnified many times over when the notion of Internet of Things (IoT) becomes a reality, i.e. objects, smart devices, services, sensors, and so on, that can interact with the user and among themselves, to provide a services or information. IoT is a service oriented architecture with resource constraints and is a mandatory subset of future Internet where every virtual and physical device can communicate with every other device giving seamless service to all stakeholders. IoT is a convergence of resource-constrained sensors, RFID, smart devices and any object with sensing, computing and communication capability. These interactions will further extend the need for privacy and security models to include how users interact with things, and how these interact among themselves. The notion of IoT and resultant business opportunities for companies across many verticals will become an important factor for tomorrow's business environment.

This chapter gives an understanding of the IoT, its vision and the economic significance. In the sequel, technical building blocks of the IoT are described

2 Internet of Things Overview

with the proposed layered architecture. We introduce the role of Radio Frequency Identification (RFID) in the IoT from a visionary technological perspective and illustrate its expected business relevance. We also describe design issues, technological and security challenges to understand how solutions to multiple problem areas in the IoT are to be designed. Finally this chapter concludes with the different IoT application areas like manufacturing, logistic relays, energy and utilities, intelligent transport, environmental monitoring and home management. We also introduce example usage scenarios to understand how different stakeholders across different business verticals will benefit from IoT.

1.1.1 Internet of Things: Vision

The IoT describes a global network of interconnecting and intercommunicating nomadic devices. It is a converged network which integrates ubiquitous computing, pervasive computing with the ambience intelligence imparted. The term IoT has different meanings for different people. In this book we use the term to mean an Internet of Things which includes objects, smart devices, services, sensors etc that can interact with the user and among themselves, using different communication methods, to provide a service or information. In such a world, the greater scale and scope of IoT increases the options in which a user can interact with the things in his environment (both physical and virtual). Efficient and novel use of technologies such as RFID [1], NFC, ZigBee and Bluetooth are contributing to create value added applications for all stakeholders in the IoT.

The main actors and the major concerns in this world are captured in the Figure 1.1 below.

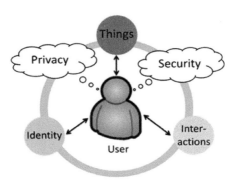

Figure 1.1 The big picture

1.1 Overview

The IoT vision can be depicted with the scenario given below.

Mark is technophile and by profession a salesman. His job requires business travels across the globe. He can access information and services both private and professional through his latest "things" developed for the future IoT. One of Mark's business trips:

At the airport

Enters the airport, gets an alert on his smart device showing the different services available at the airport (**discovery of public things and services**), e.g., a guided map of the airport, the current waiting times in the security check area, airline services etc. Chooses to check the current waiting time and is informed by a thing in the airport that on average it takes half an hour to clear security (**service consumption from public things**). At the check-in desk, another alert informs him that due to technical snag his flight is delayed by couple of hours and lunch e-vouchers are provided by airlines, (**alerts based on personal information made available to the airline thing**). Mark checks his email. The company device can access services subscribed by his company worldwide. A fast internet connection to access services and the office is available with the company's subscription (**consuming services when being part of a group**).

At the destination

It is new and unfamiliar surroundings of city for Mark and he needs to find the best route. His device discovers the local map service of the airport offered by a guide "thing" at the airport. Request for guidance to the nearest bus/train stop. Service/thing requires user's current location (access control). Mark approves for a predetermined time. The service shows the current schedule and offers the option to buy a ticket on the device (**access control to approve payment**). In the city centre, he rents a device from a local ISP for his duration of city visit (local device cheaper, customized local services (e.g. restaurants, tourist attractions, etc)). The new device is added to his "virtual terminal" (Identity imprinting/Multi device Single Sign On (SSO)) and "policies" are transferred (privacy, ID provider info, etc.) (**distributed policy management**). After the business appointment, in the city centre, he rents a bike by using a token available from a local device (payment booked via the account). The tariff depends on data from the sensors on the bike. The sensor can store the route taken and are then transferred to the local device. The route can be published on his blog, (**data Management in the sensor**). He uses the guide feature on the device. He is able to discover new "things" that offer shopping information/discounts, trips to local attractions from different ISPs, etc. He gets a proximity alert from a friend who is visiting too (Update from online location service/location based policies/etc.). They meet up for lunch. Mark allows this request on the condition that Jack presents a temporary insurance contract to use the car **(policy negotiation)** Jack negotiates a short term contract with his insurance provider and is able to provide Mark's car the proper credentials to open it and drive away (delegation mechanism). The car automatically keeps track of the user (Jack) and the fees for the tolls, parking tickets, etc. collected by him when using the car. This is automatically diverted to Jack's account instead of Mark's **(intelligent Context Management)**. Mark has lost his passport. At a police station, he provides access to his "online vault" which has the copies of his passport to the police (Distributed data access/access control). The police certify that he has lost his passport and issue a "virtual document". The "audit data" collected by the local device is uploaded to a service in the cloud. Mark's blood pressure is high and his local body sensors alert him about a possible emergency **(monitoring)**. The local health team access his body sensors so that they can retrieve the measurements. Mark accesses the Internet through the Bedside Multimedia Device which can let the home physician access latest health report and data. With help of medication his blood pressure is lowered. The collected data is automatically sent to Mark's physician for analysis.

Back at home

Mark presents the "police report" to the authorities to get a new passport issued **(secure assertions)**. He uses an audit service in the cloud to see the trip data and publishes it on his blog **(data sharing)**. An appointment is automatically scheduled with the physician.

1.1.2 Emerging Trends

The seamless interaction of nomadic devices, sensors, rooms, machines, vehicles and other devices with sensing, computing and communication capability through wired or wireless networks will add value and transform economic and technological value to future applications. This trend is becoming more visible and useful in the emerging area of entertainment as music, video and media become increasingly accessible in digital form. Any 'thing' with sensing, communication and computation capability helps us to realize the

IoT vision and there are many application areas possible due to these smart thing or objects. The stakeholders in the IoT will be [2]

- User – represents devices or software with the aim of utilization of services, infrastructure
- Providers – of services and infrastructure with the target of business
- Society – includes legal framework

These stakeholders and applications together have immediate relevance for intelligent transport system, automotive, manufacturing and logistics, retail and public security. Comapnies can utilize resulting IoT to manage all the functions like production monitoring and control, distribution, transportation and recycling of their product more effectively and efficiently. Existing paper-based system with manual operations is more prone to errors and current software systems do not have adequate and correct information that is needed to take appropriate and relevant decision in given context.

Today, the best illustration of IoT prototype can be seen with use of object identification technologies such as barcodes, NFC and RFID. RFID is the most important key enabling technologies with proven standardization protocols and set of security suites. RFID tags store unique serialized identifiers of tagged objects and it enable continuous tracking and tracing of tagged objects and as a sequel it helps to localize the tagged objects in automated mannner [3]. The location of a RFID reader can be used to determine the location of a just read RFID tag. This in turn can then be used to understand whether the "RFID touch" operation shall be used to open a door, to make a mobile payment, or to simply update a database of sightings. In future emerging RFID technology will also be in place for business process optimization [4]. All the products in the future will be equipped with sensors and actuators making them smart. This will help product and their customers to get more precise information and their surroundings. Combined with the pervasive and ubiquitous wireless communication infrastructure, mounted, embedded and attached device/object can provide information for various kinds of applications. As the today's smart devices are imparted with more smartness and equipped with increasing computing capabilities, these devices will work as centralized software server than individual host/client [5]. In the envisaged IoT, a user might want to send a document from his/her PDA to a public printer directly via a transient, peer to peer Bluetooth radio link without gaining access to a centrally administered intranet. In such ad-hoc interactions, the participating devices do not always have membership within a network. Each device will have to assume that arbitrary device can establish direct, ad-hoc communication with it. The device may simultaneously provide services to more than one network.

Some of the relevant trends in development and deployment of envisaged IoT applications can be summarized as below:

- Miniaturization of devices: Due to advancements in VLSI technology and revolution in micro and nano electronics, it is possible to develop sensors and smart devices of smaller size. This makes it easy to integrate these small devices even in living system as well as utilities.
- Mobile phones as Information gathering: Camera, NFC readers and Bluetooth makes today smart devices capable for diverse data capture and become pervasive due to the Internet connectivity. For example there are mobile applications which uses camera to read 2-D barcodes and LCD screen to display barcodes such that it can be interfaced to registration/payment systems.
- Low power devices: As the IoT devices/objects are resource constrained, energy consumption is more due to ubiquitous interactions and nomadicity. There is lot of ongoing research to contribute in this area like solar cells, powering from RF beams and energy from shakes and wind.
- Support for Big data: Current network and data management strategies/techniques are inadequate to deal with exponential traffic. Cognitive computing is playing important role in managing Big data as there will be thousand exabytes of data will be generated and stored in short future.
- Smart management: Due to economics of scale and dynamic network topology, creation of new services and applications become unmanageable. In order to cope up with this problem, envisaged IoT needs self-management, self-healing and self-configuration capability. Cooperative communication, concurrent and parallel intelligence along efficient business intelligence techniques are appropriate candidates in this direction.

1.1.3 Economic Significance

Fleisch [6] identifies economic relevance as well as key traits of the IoT that keep it apart from the current Internet:

- It will be device/service-centric than user-centric
- It will mainly consist of resource constrained devices
- Number of devices will be large in numbers: there will be trillions of network nodes than billions

- There will be communication on narrow bandwidth with resource constrained devices
- It will require lightweight protocol suite and addressing scheme

There are many value drivers in the context of machine-to-machine as well as user-to-user communication. Value drivers in machine-to-machine communication includes manual proximity trigger (check out in library, access control in building, basic payment procedure), automatic proximity trigger (supply chain management, alarm system for non-payment by customer in shop) and automatic sensor triggering (temperature sensing, humidity sensing and automatic product security (anti-counterfeiting, product pedigree). Similarly value drivers in user-to-user communication includes simple and direct user feedback (pallet identification by gate, kanban card sending and receiving wirelessly), extensive user feedback (smart thing connected to Internet, augmentation application) and mind-changing feedback (gaming console trying to kill monsters, online social networks). These value drivers are a result of a basic principle of the IoT. Due to the wide range of applications, enterprise and government are taking key interest in the development of IoT. The IoT is an integral part of emerging ICT industries and play important role in strategic alliance between ICT and non-ICT industries across the globe. Software vendors in the ICT industry have numerous opportunities for IoT. Global IoT market and all its stakeholders are predicted to increase significantly from todays $2 billion to more than 7 to 8 times by 2014. This will achieve annual significant growth of almost 50% [7].

The IoT will be key leader to achieve efficiency gain, essentially in the retail, manufacturing, logistics, energy verticals. In the highly developing countries like Europe and US, the IoT era is producing novel applications like smart home, safe and independent life for senior citizens, intelligent transport system, and environmental monitoring. This will lead towards huge job opportunities for highly qualified employees as well as highly skilled people and also will create good business opportunities for start ups. It can be envisaged that industry sectors like automotive, manufacturing and logistics, utilities, public security, and retail will undergo the big transformations in the coming years.

Expected business benefits of the IoT are summarized in [8] are as follows:

- Improvement in performance and scalability of the business processes and trends.
- Enhanced and cost-effective service provisioning through real time Big data.

8 *Internet of Things Overview*

- Transparency in localization and tracking of any kind of objects and including products and company assets.
- Potential to create new business opportunities with increased level of efficiency, accuracy, mobility, and automation.
- Improved quality of life by establishing new services and applications towards social cause.

1.2 Technical Building Blocks

The fundamental idea is that IoT will connect all objects around us to provide seamless communication and contextual services offered by them. IoT is defined as a service-oriented, distributed, and multi-layer network, and a mandatory subset of future Internet where every virtual or physical object can communicate with every other object giving seamless service to all stakeholders. IoT is a network of things which includes objects, smart devices, services, and sensors that can interact with the user, and among themselves, using different communication methods, to provide a service or information. Different applications and functional blocks of the IoT are depicted in Figure 1.2.

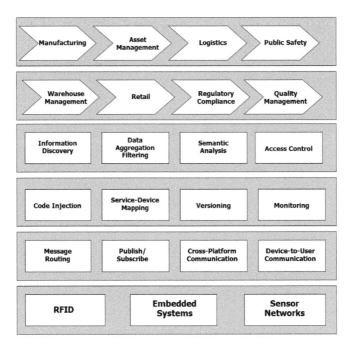

Figure 1.2 IoT applications and functional blocks

The main features of the future IoT are explained below in the points a–c.

a. *Diverse devices*

IoT includes a wide array of things, both virtual and real, ranging from smart devices with very high computing, and communication capabilities to simple sensors that give out only one piece of data (e.g. temperature sensors). Within this range, there are things like online services, virtual objects of the user placed in the network, everyday devices like cars, sensors in the house and the road, communication access points, information broadcasting devices at tourist spots, etc.

b. *Identities*

Identities are the windows through which users interact with their devices and, consume services in today's world. Before any service is delivered, it is customary to verify a digital identity of the user requesting that service (user identity), and also the identity of the entity offering the service (service identity). In IoT, this concept of identity extends to things. Ensuring that the devices have a means to be identified is critical to assure users that their interactions with the devices are safe. The identities present in the devices are also critical to their collaborative interworking.

c. *Interactions*

The ubiquitous nature of the devices will hugely impact the way in which users will interact with them in their daily life. Compared to today's world where interactions with devices, and services are restricted by ownership and, subscription (with very few exceptions), in IoT, users will be able to discover, and use things that are public. They can add things temporarily to their personal space, share their things with others, things that are public can be a part of the personal space of multiple users at the same time, etc. Such interactions require that the information shared by the user with the devices and, by devices among themselves are secure, and ensure that the authentication and, access control is preserved at all times.

1.2.1 Internet of Things Layered Architecture

The IoT networks are basically divided into three abstract layers as:
   a) ***Things:*** IoT includes a wide array of things, both virtual and real, ranging from smart devices with very high computing, and communication capabilities to simple sensors that give out only one piece of data (e.g.

temperature sensors). Within this range, there are things like online services, virtual objects of the user placed in the network, everyday devices like cars, sensors in the house and the road, communication access points, information broadcasting devices at tourist spots, etc. This layer includes all the diverse devices ranging from sensor nodes, devices with RFID tags or any other device with sensing, communication and, computing capability. These endpoints provides information storage, information collection, information processing, communication/transmission, and performance of actions.

These endpoints need unique identifiers or names for unique identification and tracking. This will help to get and manage appropriate data for related objects and map this data to appropriate application/service. A promising candidate for addressing in the context of IoT is Context-aware Clustering and Hierarchical Addressing (CCHA) [9] and for logistic scenarios, identifier like Serialized Global Trend Identifier (SGTIN) [10] is also a good option. Application scenarios where objects/devices are directly involved in communication and these devices are resource full devices (Sufficient memory and processing), IPV6 [11] is appropriate candidate as an addressing and communication protocol.

b) **Middleware:** Thing layer consist of heterogeneous devices/objects and direct communication and data exchange due to heterogeneity is difficult. Due to lack of common addressing and identification scheme, this layer also have to make the provision of maintaining the pool of things addresses and address mapping mechanism accordingly. Middleware also provides an application programming interfaces for data connectivity, directory services, event handling mechanism and security protocols. Objective of a middleware is also to abstract out the differences in data formats, addressing and configuration of underlying layers. Some of the autonomic networking features will be driven from here. This layer provides a medium for storage and, computing for aggregated information, and also provides tools for performing computations.

c) **Service/Access:** This layer is concerned with the set of techniques for accessing set of services on diverse platforms and, environments. Enterprise business applications will use and augment the information coming from device/thing layer through middleware. A piece of information (like sensor values or device identifiers) sent by sensor network or

1.2 Technical Building Blocks 11

RFID network need special provisioning and handling. Archived data for various applications will also be useful for better business planning and monitoring. Variety of applications requiring the data access and visualizations are taken care at the access layer through standard application development frameworks, and libraries. Access layer represents a set of applications to access services from IoT networks as shown in Figure 1.3.

High level IoT network architecture can be viewed as a layered architecture in which there is edge technology layer, access gateway layer, and the Internet layer. Access gateway layer consist of a collection of network devices, and gateway devices which provide connectivity between edge layer, and the Internet layer. Internet layer provides the support of Internet protocol for networking, and management.

Figure 1.3 presents a high level architecture of IoT with the functionalities of each layer [12].

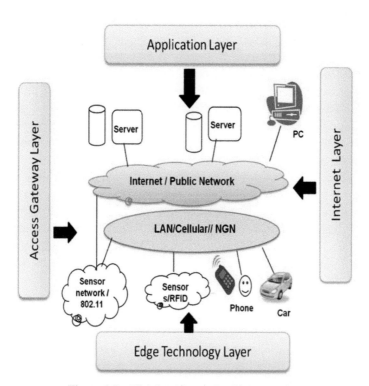

Figure 1.3 High level layered architecture of IoT

1.2.2 RFID and Internet of Things

In the ubiquitous network environment like IoT, RFID has vital role in IoT and its value proposition makes it clear that RFID will be a growing part of the web of identity that is emerging. RFID is the wireless identification technique capable of reading without direct line of sight contact through a RFID reader. RFID is a communication technology which allows for defining some unique characteristics of an object or a living being, usually its identification information, by relating it to a numeric serial number within a tag, and ensures that this number is conveyed by using radio waves. RFID provides a communication infrastructure at the radio frequencies between a special tag and reader device that can detect the tag, and allows for establishing communication between devices within the system without any physical contact, or even without seeing each other. In this regard, communication can be provided with RFID technologies in environments where technologies which require direct line of sight like barcode systems cannot be used.

An RFID tag is made up of an RFID chip attached to an antenna. Tags can be read from several meters away and beyond the line of sight of the reader. The most common method of identification is to store a serial number that identifies a person or an object, but an RFID tag can also store some additional information depending on the size of its memory. A tag is attached physically to the object to be identified. The tag is an electrical device designed to receive a specific signal and automatically transmit a specific reply. Tags can be passive or active, based on their power source and the way they are used, and can be read-only, read/write or read/write/re-write, depending on how their data is encoded. Passive RFID tags take the energy from the electro-magnetic field emitted by readers. Tags use the transmitting frequency in kilohertz, megahertz, and gigahertz ranges. As IoT provides two important basic functions for an Internet of Things – identification and communication – RFID can also be used to determine the approximate location of objects provided the position of the reader is known. Figure 1.4 shows communication between tags and reader.

As the RFID is integral and mandatory part of IoT for object identification in various contexts, in the long term, infrastructure such as the EPC network will play an important role [13]. The EPCglobal Network [13] is interface standard for identity capture and exchange. Electronic Product Code (EPC) [14] is widely accepted and emerging solution for Radio Frequency Identification (RFID) for the identification and tracking applications. EPC is

1.2 Technical Building Blocks 13

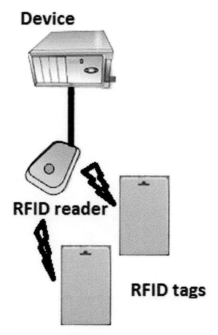

Figure 1.4 RFID tag-reader communication

a numeric attribute assigned to any RFID tag and represents unique product code with significant characterization and categorization for precise meaning. EPCGlobal have defined a set of standards that consists of collection of hardware, software, and data standards, to enable the accurate and effective operation especially throughout the supply chain process using RFID [14]. Figure 1.5 depicts the overall view of EPCGlobal Architecture Framework and the interaction among the components within and inter-enterprise. As we can see, it consists of specifications of components down from the RFID tag and reader level, up to the high-level data and information services. In this section, we will describe more specifically on the EPCIS and ONS as the relevant entities to the inter-connection of RFID network over the global network.

1.2.2.1 EPCIS

EPCIS is network service to provide EPC specific RFID enabled product information. ONS is only maintained by manufacturers and EPCIS may be deployed by all parties who need EPC of tagged product. There are 3 main

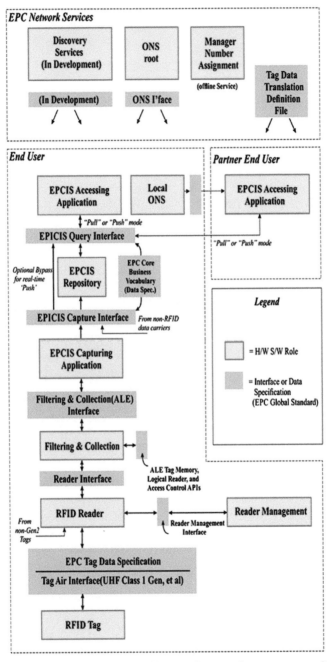

Figure 1.5 EPCGlobal overall set of architecture framework components and layers [14]

parts of EPCIS that are defined in the EPCGlobal specification namely EPCIS capture interface, EPCIS repository, and EPCIS query interface.

- EPCIS capture interface is defined as an interface that is responsible to provide a path for communicating EPCIS events generated by EPCIS Capturing Applications to other roles that require them, including EPCIS Repositories, internal EPCIS Accessing Applications, and Partner EPCIS Accessing Applications. In the actual RFID Middleware, EPCIS capturing application that interprets the captured RFID data from the lower layer of Middleware, e.g. Filtering & Collection (F&C) layer or other component depending on the implementation needs to be implemented along with the EPCIS capture interface.
- EPCIS repository is a database system that provides persistence information not only raw RFID data, but also other RFID related events which are defined in F &C layer as well as business context information.
- Those data can then be accessed by accessing application, both internal and external, through EPCIS query interface as defined in EPCGlobal specification.

1.2.2.2 ONS

- The Object Name Service (ONS) is a service that returns a list of network accessible service endpoints that pertain to a requested EPC. The ONS does not contain actual data about the EPC; it only contains the network address of services that contain the actual data [15]. The ONS uses the Internet's existing DNS for resolving requests about an EPC. In order to use DNS to find information about an item, the item's EPC must be converted into a format that DNS can understand, which is the typical, "dot" delimited, left to right form of all domain-names. The ONS resolution process requires that the EPC being asked about is in its pure identity URI form as defined by the EPCglobal Tag Data Standard [14]. This URI conversion is done in the local server by the TDT component.
- The key components of the logical architecture are the Core ONS servers. The core of the system comprises by the ONS servers themselves. Each company assigned part of the global EPC namespace is responsible to create and maintain a functional ONS server to serve queries about EPC's inside that namespace. Each ONS server is basically a standard DNS server with special NAPTR records [ONS], mapping EPC's to service access points – typically EPCIS query and capture

interfaces where additional information can be obtained about the EPC in question [15].

Discovery of object with ONS and EPC discovery service and EPC security framework for secure access are main component of EPCglobal Network. RFID system consist of transponder i.e. tag itself and transceiver, i.e. reader and to track any object, RFID uses EPC. RFID reader reads EPC from tag and IP based local system collects information pointed by EPC ONS protocol. EPCIS (EPC Information Service) servers' process ONS query and returns PML (Physical Markup Language) files to present meaningful information related to tagged object. Match is performed at ONS server between EPC number stored on the tag and the address of EPCIS server and reply is given to local system where to find information about object. RFID system architecture is shown in Figure 1.6. ONS works as a directory service that routes request for information about EPC between requesting party and product manufacturer. ONS is lookup service which takes EPC as an input and outputs the address of EPCIS service in the form of Uniform Address Locator (URL). ONS architecture is similar to the Internet Domain Name System (DNS) but in DNS lookup services are implemented hierarchically. ONS works as follows:

- When end user application or client needs to find EPCIS service, it queries to local DNS resolver.
- Local DNS performs multi-step lookup to return the result to requesting application.
- It consults to ROOT ONS of EPCglobal
- ROOT ONS identifies the LOCAL ONS
- LOCAL ONS provides pointer to the appropriate EPCIS server

In nutshell, ONS maps EPC represented with URI form of the EPC tag ID into IP address. Root ONS only keeps the IP address of the target local ONS, so when a client send a query with a particular EPC ID, the root ONS returns IP address of target local ONS to the client, then the client will further query the given local ONS and finally get the IP address of the corresponding EPCIS. Just like DNS that returns the IP address of the server which hosts the website.

1.2.3 IP for Things

IP provides protocol for implementing IoT applications. Integrating RFID network component with current IP network using Domain Name System

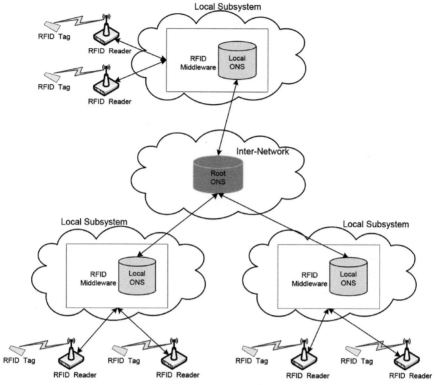
Figure 1.6 RFID system architecture

(DNS) [16, 17] and Dynamic Host Configuration Protocol (DHCP) [18] is the most efficient way. However this integration introduces problem of authentication, authorization and confidentiality. Using security extension of DNS aforementioned security issues can be addressed. In IoT, thing of interest has one identity but can be identified by several different identifiers associated with it. These identifiers are used to distinguish two things from each other and depending on the profile different identifier might be used. DNS works as a most important infrastructure for internet naming service. DNS is TCP/IP based distributed name service to map host names to IP addresses. IoT needs name service similar to DNS to map things names in heterogeneous namespace to corresponding IoT addresses. Due to mobility, dynamic nature, heterogeneous namespace and resource constraints in IoT, IoT name service needs special attention. Due to scalability of IoT, name to address mapping needs to be efficient in time. Efficiency and compatibility issues impose two design parameters on IoT name service which are lightweight and heterogeneous. In IoT, integrating things capabilities and interconnecting small mobile energy constrained sensor and RFID tags is critical. In RFID based identification scheme, it allows each thing to have unique identifier which is readable at distance supporting automatic and real time identification of things.

Advantages of integrating IoT with IP are:

- Support for heterogonous communication standard and interoperability
- Applicable to transport protocol for end-to-end reliability
- Well established DNS based naming and addressing and lookup services
- Application level protocol support like HTTP and XML
- Good network management support like SNMP

1.3 Issues and Challenges

The IoT is advanced network which includes physical objects together with powerful computers and other smart devices. IoT comprises ad-hoc networks with the collaborative capabilities due to varied types of objects which can be classified depending on different parameters. Few parameters for classification are size, mobility, power, connectivity, automation and network protocol. The IoT objects have characteristics like sensing and/or actuating ability, energy/power limitation, mobility, and the connectivity with the physical world. With respect to classification and characteristics of the IoT objects, different issues and challenges are described in the section below.

1.3.1 Design Issues

Seamless integration of the 'things' to the internet will be challenging. The vision of IoT is to connect every object with computing, communication and sensing ability to the Internet. IoT contains varied range of devices from RFID tags, sensor nodes to the even shoes. Major factors of influence are the connectivity, power sources, form factor, security, geographical factors and cost of deployment and operation. Looking into business applications of the IoT, it is envisaged that scalability, modularity, extensibility and interoperability are key design requirements for the IoT, in order to provide services to the solution providers and developers and the users. In the IoT the range of connectivity options will increase gradually and in this context the communication needs will change and new radio and service architectures

1.3 Issues and Challenges

will be required to cater for the connectivity demands of emerging devices. Figure 1.7 summarizes key factors to be considered for the IoT. These design constraints shall play an important role in designing infrastructure and protocol for the IoT.

- **Low memory:** IoT objects are equipped with low memory (E.g, sensor nodes, RFID tags)
- **Low computational power:** Objects have very low computing power which can process a small piece of data.
- **Low life time:** Due to seamless and nomadic service provision to the users, power supply is another important issue.
- **Low bit rate and throughput:** Depending on the underlying application, amount of data generated is varied, and due to the scale of economics, massive data is generated. This is one of the important challenges in resource constrained IoT.
- **Bandwidth:** IoT comprises of small scale (Smart Home) or large scale application (Factory or Mall) area offering seamless services to users,

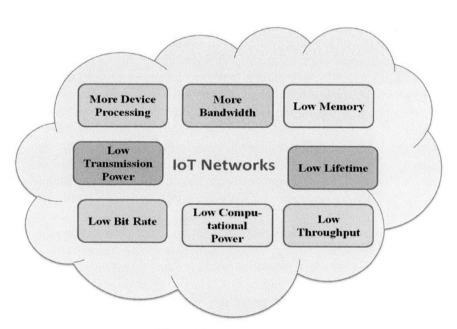

Figure 1.7 IoT design issues

and other devices. Scalability is an important issue to uniquely identify individual devices and results into more bandwidth requirement.

It is clear that we are transforming from an Internet of computers to the IoT with device-to-device communication. In order to make the IoT services available at low cost with a large number of devices communicating to each other, there are many challenges to overcome. These challenges are divided into two categories as:

a. Technological challenges
b. Security challenges

1.3.2 Technological Challenges

These challenges are related to underlined wireless technologies, energy, scalability, distributed and dynamic nature of IoT and ubiquitous interactions.

- **Wireless Communication:** IoT significantly uses convergence of established wireless technologies such as GSM, UMTS, Wi-Fi, Bluetooth and WPAN. These underlined wireless technologies use different standards and have different communication bandwidth requirement. This convergence also creates serious interoperability issues.
- **Scalability:** Unbounded number of devices creates the larger scope and scalability in IoT than conventional communication networks. IoT covers large application areas like a home environment where number of devices are relatively small in number to a factory or building that has a large number of devices offering multiple services to the users. IPV6 is one attempt to accommodate as many numbers of devices and things in IoT.
- **Energy:** IoT consist of constrained objects which do not have enough power, memory and computation capabilities. Designing lightweight protocols for IoT which minimize energy consumption is very important as compared to conventional protocols running on devices with sufficient resources.
- **Distributed and Dynamic Nature:** In IoT, things can interact with other things at any time, from anywhere and in any way independent of the location. As the IoT networks are distributed in nature, designing protocols for them is a challenging task. The objects interact dynamically and hence appropriate services for the objects must be automatically identified. In addition to this, the mobility/roaming of the objects is another important challenge.

- **Identification:** In the IoT, things include variety of objects like computers, sensor nodes, people, vehicles, medicines, books, etc. These things should be uniquely identified for the addressing capabilities and for providing a means to communicate with each other. After verifying the identities of things, we call these uniquely identified things as objects. Different identity schemes have been proposed for the IoT and it is predicted that it is dubious to have common identification schemes globally. Identification schemes like RFID Object Identifier, EPCglobal, Short-OID and Near Field Communications Forum, IPV4, IPV6 and E.164 have been studied in the literature. These addressing methods/principles are highly depends on the underlined access technology, thus it is challenging to have many different addressing protocols for varied underline access technologies.

1.3.3 Security Challenges

These challenges are related to security services like authentication, privacy, trustworthiness and confidentiality. Security challenges also include heterogeneous communication and end-to-end security.

- **Privacy:** Privacy is one of the most sensitive areas in the context of IoT. In IoT, all objects are connected to the Internet and they communicate with each other over the Internet. Hence the privacy issue is critical. As the Internet gets diversified with new types of devices and heterogeneous networks, IoT users and devices have to access the digital world with wide range of methods and protocols. Further, as ownership of these devices by the users does not exist, the issue of privacy is aggravated.
- **Identity Management**: Due to the scale of economics in the IoT, unbounded numbers of things or objects are involved in accessing IoT networks and communicating with each other. Hence, efficient and lightweight identity management schemes are required. In addition to this, the distributed nature of IoT makes this problem more challenging.
- **Trust:** Trust is an essential and integral factor to consider when implementing IoT. In an uncertain IoT environment, trust plays an important role in establishing secure communication between things. There should be an effective mechanism to define trust in a dynamic and collaborative IoT environment. It is also important to provide context aware trust management for varied IoT applications.
- **End-to-End Security:** End-to-end security measures between IoT devices and Internet hosts are equally important. Applying cryptographic

schemes for encryption and authentication codes to a packet is not sufficient for the resource constrained IoT. Hence future research is required into efficient end-to-end security measures between IoT and the Internet.
- ***Authentication and Access Control:*** Authentication is identity establishment between communicating parties. Authentication and access control is important to establish secure communication between multiple devices and services. Interoperability and backward compatibility are the two key issues to be addressed. For example, in Wi-Fi roaming, devices use UMTS at the core networks.
- ***Attack Resistant Security Solution:*** Due to diversity of devices and end users, there should be attack resistant and lightweight security solutions. All the devices in IoT have low memory and limited computation resources, thus they are vulnerable to resource enervation attack. When the devices join and commissioned into the network, keying material, security and domain parameters could be eavesdropped. Possible external attacks like denial of service attack, flood attack, etc., on device and mitigation plan to address these attacks is another big challenge.

1.4 Applications

There is a wide range of applications of IoT and these IoT applications are categorized in four domains in [19] as
- Personal and Home – Includes individual home [20, 21]
- Enterprise – Includes scale of community [22, 23]
- Utilities – Includes national and regional scale [24, 25]
- Mobile – Includes IoT applications spread across multi-domain due to distributed connectivity and scale [26, 27]

Major application areas include the following [28, 29]:

1.4.1 Manufacturing, Logistic and Relay

Deploying IoT for supply chain management can leverage many advantages. Items or goods equipped with RFID tags, retailer can optimize many activities like automation in checking, effective and real time stock monitoring, detecting expired stock. Furthermore, this on demand information regarding goods can also optimize the logistics of whole supply chain. Applications also include authentication of goods, anti-counterfeiting, inventory management, service and support.

1.4.2 Energy and Utilities

As the number of consumers and users are increasing at faster rate, IoT will play a key role in energy and utilities sector. It includes smart electricity grid and water transmission grid, real time monitoring of water supply and electricity usage. There are two fronts in this area like service provider and consumer. At the consumer end, efficient utilization of electricity and water at home enabled by smart devices to the grid is important emerging application area.

1.4.3 Intelligent Transport

Use of sensor network in vehicular ad-hoc networks, GPS and wireless network is increasing at faster rate. Use of GPS for localization and tracking, and its convergence will make the vehicles and transportation system smart. Vehicle-to-vehicle communication and vehicle to hotspot communication and its integration with the Internet will enable road safety and efficiency. Vehicle tracking, traffic data collection for management, traffic rule enforcement systems, automotive infotainment systems are going to be a part of an integrated network. Detection of real time traffic and use of video sensors connected to the Internet for traffic forecasting is another emerging application.

1.4.4 Environmental Monitoring

Whether forecasting and environmental monitoring is very important and valuable application connected to the agricultural sector. It includes the use of sensor network where sensor nodes are used for monitoring environmental parameters, soil conditions. This will have extensive use in agriculture, security, surveillance and disaster management. These parameters can also be used for weather forecasting.

1.4.5 Home Management

This scenario presents a way for optimizing home services. The envisioned homes of the future will mainly consist of places full of things that will interact with each other at different levels. We will encounter different kinds of sensors and devices that might use heterogeneous technologies: low bandwidth mesh networking based (such as Insteon, ZigBee and Z-Wave) or other more bandwidth demanding (such as Bluetooth, Wi-Fi, 4G or UWB) providing 24x7 monitoring or entertainment services. The result of this data gathering will be used to trigger different user defined alarms that will be centralized in one or more mobile devices, such as the parent's mobile phones or the home TV, depending on the current conditions. The access to this data and to the all available devices is to be ubiquitously granted by all entities allowed by the enforced access control policies.

Summing up, the main goals of this scenario are:

- Ubiquitous access to services or monitoring data granted to Identities that fulfil the access policies.
- Alarm triggering and monitoring centralized in mobile devices.
- Heterogeneous device interaction.

1.4.6 eHealth

One of the most important scenario where IoT (sensors, actuators, RFID tags, etc.) is planned to be used and being applied is eHealth. The main objective is to provide ease of life including health services across geographic and time barriers. eHealth scenario will allow tele-monitoring of the environment and health conditions of a person may it be chronic or by accident, while at home or abroad. Especially in the case of user is travelling to a foreign destination, to obtain access to the medical history and record of the patient becomes a critical issue in order to establish the right diagnostic, by emergency services or hospital. This puts privacy as very important research criteria in order to keep non-authorized people from accessing the medical and user information.

Summing up:

- Remote medical monitoring.
- Access to medical history and electronic patient records from anywhere.
- Use of IoT in eHealth

1.5 Conclusions

One of the most profound changes today is the increase in mobility of portable yet powerful wireless devices capable of communicating via several different kinds of wireless radio networks of varying link-level characteristics. In the last few years the IoT has seen widespread application and can be found in each field. In the IoT, every virtual and physical entity is communicable, addressable and is accessible through the Internet. These virtual and physical entities produce seamless communication and seamless service collaborating with users and other devices creating service oriented networks. The IoT is an emerging paradigm and makes the world of computing fully ubiquitous creating UbiComp, a term initially coined by Mark Weiser [30]. Any "thing" with sensing, communication and computation capability helps us to realize the IoT vision and there are many application areas possible due to these smart thing or objects.

Securing user interactions with IoT is also essential if the notion of "things everywhere" is to succeed. Mobility is very important aspect of mobile and wireless communication and essentially in the context of IoT. With the heterogeneous network topologies like Wi-Fi, LTE and WiMax, authenticated service delivery with proper access control in place on the fly is a big challenge.

We have discussed the emerging trends and economic significance of the IoT in business and enterprise perspective. IoT vision is discussed by depicting the motivational scenario of envisage IoT. A discussion on RFID, IP along with the layered architecture of the futuristic IoT is then elaborated as a technical building blocks.

The main features of the future IoT are explained along with the design issues. Clear understanding of these design issues is important in order to design protocols or security solution for the IoT. It is clear that we are transforming from an Internet of computers to the IoT with device-to-device communication. In order to make the IoT services available at low cost with a large number of devices communicating to each other, there are many challenges to overcome. In the sequel, technological and security challenges are presented and discussed. Finally a discussion on the wide range of applications which includes supply chain management, utilities, intelligent transport, environmental monitoring and home management concludes this chapter.

References

[1] K. Finkenzeller, RFID Handbook: Fundamentals and Applications in Contactless Smart Cards and Identification, 2nd Edition, John Wiley and Sons, 2003.

[2] Amardeo Sarma, and Joao Girao, "Identities in the Future Internet of Things," In Springer Wireless Personal Communications, Volume: 49, Issue: 3: pp: 353–363. May 2009.

[3] T. Gotz, S. Safai, Ph.Beer, "Efficient supply chain management with SAP solutions for RFID," Galielo Press New York, 2006.

[4] L. Yan, Y. Zhang, L. T. Yang, H. Ning (Eds), "The Internet of Things: From RFID to the Next Generation Pervasive Networked Systems," Auerbach Publication, 2008.
[5] M. Marin-Perianu, N. Meratnia, P. Havinga, L. Moreira Sa de Souza, J. Muller, P. Spiess, S. Haller, T. Riedel, Ch. Decker, G. Stromberg, "Decentralized Enterprise System: A Multiplatform Wireless Sensor Approach," In IEEE Wireless Communications, Vol. 14, December 2007.
[6] E. Fleisch, "What Is the Internet of Things? An Economic Perspective," white paper, WP-BIZAPP-053, AutoID Labs, Jan. 2010; www.autoidlabs.org.
[7] ABI Research: RFID Market Update, 2006.
[8] ICT shaping the world: A scientific View © 2009 ETSI.
[9] Parikshit N. Mahalle, Neeli R. Prasad and Ramjee Prasad, "Novel Context-aware Clustering with Hierarchical Addressing (CCHA) for the Internet of Things (IoT)," In the Proceedings of IEEE Fourth International Conference on Recent Trends in Information, Telecommunication and Computing – ITC 2013, August 01–02, 2013, Chandigarh, India.
[10] EPCglobal: Tag Data Standards, Version 1.4, EPCglobal Ratified Specifications, June 2008, http://www.epcglobalinc.org/standards/tds/tds _1_ 4-standard-20080611.pdf (August 2008).
[11] S. Deering, R. Hinden, "Internet Protocol, Version (IPV6) Specification, IETF Network Working Group, RFC 2460, December 1998.
[12] Parikshit N. Mahalle, Sachin Babar, Neeli R Prasad, and Ramjee Prasad, "Identity Management Framework towards Internet of Things (IoT): Roadmap and Key Challenges," In proceedings of 3rd International Conference CNSA 2010, Book titled Recent Trends in Network Security and Applications - Communications in Computer and Information Science 2010, Springer Berlin Heidelberg, pp: 430–439, Volume: 89, Chennai-India, July 23–25 2010.
[13] Thiesse, F., Floerkemeier, C., Harrison, M., Michahelles, F., Roduner, C.: Technology, Standards, and Real-World Deployments of the EPC Network. IEEE Internet Computing 13(2): 36–43 (2009)
[14] The EPCglobal Architecture Framework Version 1.2 at http://www.epcglobalinc.org
[15] John Soldatos, et al, "Core ASPIRE Middleware Infrastructure (Final Version)," Public report – Deliverable, ASPIRE collaborative project.
[16] Mockapetris P., "Domain Names – Concepts and Facilities", RFC 1034, November 1987.

[17] Mockapetris P., "Domain Names – Implementation and Specification", RFC 1035, November 1987.
[18] R. Droms, "Dynamic Host Configuration Protocol", RFC 2131, March 1997
[19] Jayavardhana Gubbi, Rajkumar Buyya, Slaven Marusic, and Marimuthu Palaniswami, "Internet of Things (IoT): A Vision, Architectural Elements, and Future Directions," Technical Report CLOUDS-TR-2012-2, Cloud Computing, and Distributed Systems Laboratory, The University of Melbourne, June 29, 2012.
[20] Xiaodong Lin, Rongxing Lu, Xuemin Shen, Nemoto Y., and Kato N., "SAGE: A Strong Privacy-preserving Scheme Against Global Eavesdropping for ehealth Systems," In IEEE Journal on Selected Areas in Communications, volume: 27, Issue: 4, pp: 365–378. May 2009.
[21] Rohokale Vandana M., Neeli R Prasad and Ramjee Prasad, "A Cooperative Internet of Things (IoT) for Rural Healthcare Monitoring, and Control," In Proceedings of IEEE Wireless Vitae 2011, 2nd International Conference on Wireless Communications, Vehicular Technology, Information Theory, and Aerospace & Electronic Systems Technology, pp: 1–6. Chennai – India, February 28 – March 3 2011.
[22] A. Gluhak, S. Krco, M. Nati, D. Pfisterer, N. Mitton, and T. Razafindralambo, "A Survey on Facilities for Experimental Internet of Things Research," In IEEE Communication Magazine, Volume: 49, Issue: 11, pp: 58–67, November 2011.
[23] X. Li, R.X. Lu, X.H. Liang, X.M. Shen, J.M. Chen, and X.D. Lin, "Smart Community: An Internet of Things Application," In IEEE Communications Magazine, Volume: 49, Issue: 11, pp: 68–75, November 2011.
[24] O. Garcia-Morchon, "Security Considerations in the IP-Based Internet of Things," IETF, Mar. 2011; http://tools.ietf.org/html/draft-garcia-core-security.
[25] P. Spiess, S. Karnouskos, D. Guinard, D. Savio, O. Baecker, L. Souza and V. Trifa, "SOA-based Integration of the Internet of Things in Enterprise Services," In Proceedings of IEEE International conference on web services (ICWS 2009), pp: 968–975, Los Angeles, CA – USA, July 6–10 2009.
[26] I.F. Akyildiz, J. Xie, and S. Mohanty, "A Survey on Mobility Management in Next Generation All-IP based Wireless Systems," In IEEE Wireless Communications Magazine, Volume: 11, Issue: 4, pp: 16–28, August 2004.

[27] Y.W. Ma, C.F. Lai, Y.M. Huang and J.L. Chen, "Mobile RFID with IPv6 for Phone Services," In Proceedings of IEEE International Symposium on Consumer Electronics (ISCE 2009), pp: 169–170, Kyoto- Japan, May 25–18 2009.
[28] Disruptive Technologies: Global Trends 2025, SRI Consulting Business Intelligence, Appendix F: The Internet of Things, 2008.
[29] O. Vermesan, M. Harrison, H. Vogt, K. Kalaboukas, M. Tomasella et al. (Eds.), "The Internet of Things - Strategic Research Roadmap", Cluster of European Research Projects on the Internet of Things, CERP-IoT, 2009.
[30] M. Weiser, "The computer for the 21st Century," Scientific American, Volume: 265, pp: 66–75, September 1991.

2

Elements of Internet of Things Security

2.1 Introduction

IoT is a novel paradigm which is becoming popular in research community and industry due to its wide range of applications. The fundamental idea is that IoT will connect all objects around us to provide seamless communication and contextual services offered by them. Economics of scale in the IoT presents new security challenges for ubiquitous devices in terms of authentication, addressing and embedded security. Devices like RFID and sensor nodes most often have no access control functionality and can freely obtain information from each other. As a result, an authentication as well as authorization scheme must be established between these devices to achieve the security goals for IoT. Without any strong security, IoT malfunctions and attacks will overweigh of its benefits. Privacy of things and security of data is one of the key challenges in the IoT. Security will become more serious issue as the IoT becomes an integral part of everyday life. The numbers of embedded systems like refrigerator, washing machines to TV are connected to the Internet, but the vast majority of these systems are un-patchable, or poorly maintained.

Pervasive and ubiquitous nature of IoT makes a set of new challenges beyond merely making the systems work, and prominently amongst the challenges is to provide improved security. This chapter presents requirements and challenges for handling successful security in IoT. In this chapter threat modeling, threat analysis and use cases and misuse cases are also discussed.

2.1.1 Vulnerabilities of IoT

General security needs and devices life cycle in the context of IoT for Building, Automation and Control (BAC) system are presented in [1]. In BAC system, there is a network of interconnected nodes that performs various functions like heating, ventilating and air conditioning. All nodes carries different functionality and maximum of these devices are resource constrained like sensor nodes. Life of devices starts when they are manufactured to perform specific tasks and hence there are devices from different manufacturers. Due to this reason, trust bootstrapping and interoperability are major issues. Next phase is installation and commissioning within IoT network based on device identity and secret keys. Procedures for installation and bootstrapping are defined for fix period of time. After this, device become operational and runs the functions of BAC system. During operational phase, devices are under the control of resource owner and occasional maintenance is required. Maintenance includes software up-gradation and reconfiguration. Due to operational changes on devices, they may require re-bootstrap. The device continues for the operational phase and the eventual maintenance phase until the device is decommissioned at the end of its lifecycle. Figure 2.1 shows the generic lifecycle of a thing. This generic lifecycle is also applicable for the IoT scenarios other than BAC systems.

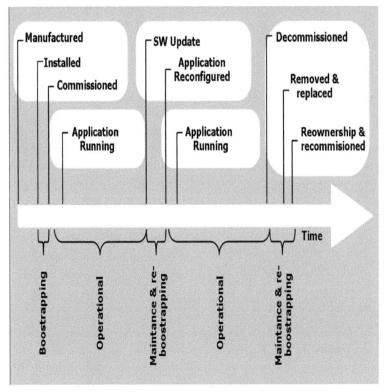

Figure 2.1 The life cycle of a device in IoT [1]

2.1 Introduction   31

Life cycle of devices shows that, there are many vulnerabilities and security relationship between devices and secure interaction need to be addressed. In the IoT context, security is not limited to the required security services, but should be also extended to overall system and functionalities. Vulnerabilities are fact of life in IoT and information security. Dynamic network topology and, distributed nature makes IoT more vulnerable to security threats, and attacks. Mobility and weak physical security of low power devices in IoT networks are also possible causes for security vulnerabilities. Attacks are grouped into two types: passive attacks and active attacks. In passive attacks, attackers are interested in eavesdropping and monitoring of data transmission. In other words, attacker does not attempt to perform modifications. Active attacks can be in the form of modification, fabrication and interruption. Denial of service (DoS) attacks is one of the example active attacks. Threats include identity theft through masquerading or spoofing, unauthorized access to resources, unauthorized disclosure or modification of data. IoT opens your home to cyber threats. With reference to device life cycle in the IoT, Figure 2.2 depicts vulnerabilities of IoT. Possible vulnerabilities of IoT are as follows:

1. **Unauthorized access:** One of the main threats is the tampering of resources by unauthorized access. These access rights may be granted to

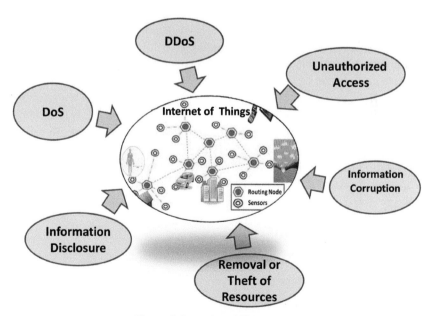

Figure 2.2   Vulnerabilities of IoT

an unauthorized entity if an attacker is able to get hold of the authorization process. Identity-based verification should be done before granting the access rights.
2. **Information corruption:** Other threat is information corruption and, to address this, the device credentials must be protected from tampering. Secure design of access rights, credential and, exchange is required to avoid corruption.
3. **Theft of resources:** The access of shared resources over insecure channel causes theft of resources, or data flow, and results into man-in-the-middle attack.
4. **Information disclosure:** In IoT, the data is stored at different places in different forms depending on the context. This distributed data must be protected from disclosure. The context-aware access control must be enforced to regulate access to system resources.
5. **DoS attack:** A DoS attack makes an attempt to prevent legitimate user from accessing services which they are eligible for [2]. For example unauthorized user sends to many requests to server so as to flood the network and deny other legitimate users from access to the network.
6. **DDoS:** Distributed Denial of Service (DDoS) is a type of DoS attack where multiple compromised systems – which are usually infected with a Trojan – are used to target a single system causing a Denial of Service (DoS) attack. Victims of a DDoS attack consist of both the end targeted system and all systems maliciously used and controlled by the hacker in the distributed attack [3].

CyberBunker Launches "World's Largest" DDoS Attack, Slows down the Entire Internet. A massive cyberattack launched by the Dutch web hosting company CyberBunker has caused global disruption of the web, slowing down internet speeds for millions of users across the world, according to a BBC report. CyberBunker launched an all-out assault, described by the BBC as the world's biggest ever cyberattack, on the self-appointed spam-fighting company Spamhaus, which maintains a blacklist used by email providers to filter out spam [4].

Here are few real examples of attacks that hit the IoT [5].
1. First there was the Carna Botnet. At its peak, 420,000 'things,' such as routers, modems, printers were compromised.

2. Then TRENDnet's connected cameras were hacked, with feeds from those cameras published online, forcing the FTC to make its first ever IoT judgement.
3. Another is the Linux.Darlloz – PoC IoT worm found in the wild by Symantec, while most recently Proofpoint discovered a Botnet of 100,000 compromised systems including connected things such as TVs, routers and even a fridge.

2.1.2 Security Requirements

IoT security requirements to counter the threats like tampering, fabrication and theft of resources are listed below:

1. *Access control*

    The access control provides authorized access to network resources. IoT is ad-hoc, and dynamic in nature. Efficient and a robust mechanism of secure access to resources must be deployed with distributed nature.

2. *Authentication*

    Authentication is an identity establishment between communicating parties (devices). Due to diversity of devices, and end users, there should be an attack resistant and lightweight solution for authentication.

3. *Data confidentiality*

    Data confidentiality is protecting data from unauthorized disclosure and data tampering. Secure, lightweight, and efficient key exchange mechanism is required due to dynamic network topology.

4. *Availability*

    Availability is ensuring no denial of authorized access to network resources. Access control and availability problems are critical due to the wireless nature of ad-hoc networks.

5. *Trust management*

    Trust management, and trust-based access control are basic requirements in IoT due to its nomadic nature. Decision rules needs to be evolved for trust management in IoT.

6. *Secure software execution*

    It refers to a secure, managed-code, runtime environment designed to protect against deviant applications.

7. *Secure storage*
   Secure storage involves confidentiality and integrity of sensitive information stored in the system.

8. *Tamper resistance*
   It refers to the desire to maintain these security requirements even when the device falls into the hands of malicious parties, and can be physically or logically probed.

9. *Scalability*
   IoT system will consist of various types of devices in terms of different capabilities (from intelligent sensors and actuators, to home appliances) as well communication means (wire or wireless) and protocols (Bluetooth, ZigBee, RFID, Wi-Fi, etc), and across different geographical locations. As a result, the system is highly distributed, heterogeneous, and pervasive. Dealing with such type of system, scalability is an important point in designing a security solution.

10. *Flexibility and adaptability*
    IoT will likely to consist of mobile communication devices which can roam around freely from one type of environment to the others with different type of risks and security threats. Furthermore, users are likely to have different privacy profile depending on environment or with whom they are communicating. Therefore, flexibility and adaptability are the other important requirements for a security solution in IoT.

Figure 2.3 depicts high level security architecture for IoT with possible threats, and attacks. This architecture provides systematic way of countering the above threats. Right side of the architecture shows possible threats in IoT. Threats include destruction of resources by unauthorized access, information disclosure, information corruption, theft of resources, and information disclosure. Security dimensions shown in this architecture are the mitigation principles to counter these threats.

As explained and presented in the Figure 2.3, main security requirements/objectives in IoT includes access control, authentication, confidentiality, availability and the trust management.

2.1.3 Challenges for Secure Internet of Things

IoT is an intelligent collaboration of tiny sensors and devices giving new challenges to security and privacy in end-to-end communication of things. Protection of data and privacy of things is one of the key challenges in the IoT [6].

2.1 Introduction

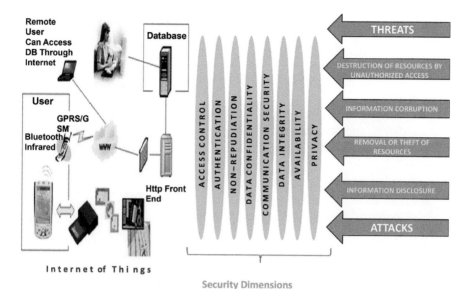

Figure 2.3 Security architecture for IoT

Security challenges identified in the IoT context are as follows:

1. **Identity management (IdM) for devices**
    - New Identity concepts, and their implications in IoT world
    - Identity delegation, imprinting of identity in things, merging identities to create a meta-identity, etc.
    - Trust Management, Circles of Trust (IoT belonging to different owners)
    - Identity and privacy
    - Authentication schemes for IoT
    - Secure attribute exchange, and selective disclosure of attributes inside IoT

2. **Secure interactions in/with IoT**
    - Secure, and certified context information for things
    - Reliable computation, and storage services provided by IoT
    - Interaction of things in a Better-Than-Nothing Security (BTNS) environment
    - Secure, and dynamic network, and space composition, discovery, namespace, resolution and indexing of things
    - Auditing of interactions with things
    - Physical and virtual mobility of things

3. **Distributed access control and privacy**
    - Dynamic exchange of authenticated identity information between things
    - Credential Management, and bootstrapping with single sign on for things
    - Privacy-aware policy-based authorization systems with deductive policies, and delegation
    - Dynamic selection of applicable policies based on the environment in IoT
    - Dynamic attributes negotiation for things
    - Proxy security services with delegation for things, in particular, for 6LowPAN devices privacy-aware negotiation, and application of attribute releasing policies

4. **Secure data management and exchange**
    - Assurance for the information exchange between things
    - Secure, and private management of distributed data spread across multiple things
    - Personal data auditing, and enhanced audit data visualization for users to make them understand the usage of their identities, and data by things
    - Signed context information for exchange with things controlled by user privacy policies
    - Secure storage, and deletion of audit data in a distributed IoT environment

5. **End-to-End security**
    - End-to-end security measures between IoT devices and Internet hosts are equally important.
    - Applying cryptographic schemes for encryption and authentication codes to a packet is not sufficient for the resource constrained IoT.
    - Hence future research is required into efficient end-to-end security measures between IoT and the Internet [7].

6. **Privacy**
    - Privacy is one of the most sensitive areas in the context of IoT.
    - In IoT, all objects are connected to the Internet and they communicate with each other over the Internet. Hence the privacy issue is crucial.

- As the Internet gets expanded with new types of devices and heterogeneous networks, IoT users and devices have to access the digital world with wide range of protocols and methods.
- Further, as ownership of these devices by the users does not exist, the issue of privacy is getting more serious [7].

7. **Security structure**
   The IoT will remain stable-persisting as a whole over time. In the sequel, putting together the security mechanism of each logical layer cannot implement the defense-in-depth of system [8]. So it is a challenging and important research area to construct security structure with the combination of control and information [9]. Challenges presented above shows that, there is a need of integrated approach of authentication, and access control for ubiquitous devices in IoT. Furthermore, the solution for authentication and access control must be attack resistant from the well-known attacks.

2.2 Threat Modeling

Threat modeling is presented by first defining misuse case i.e. negative scenario describing the ways the system should not work and then standard use case. The assets to be protected in IoT will vary with respect to every scenario case. The modeling of the security attacks helps to understand an actual view of the IoT networks and enable us to decide the mitigation plans [7].

2.2.1 Threat Analysis

We recommend that the assets needs to be identified to drive threat analysis process and also to guide specification for security requirements. Let's consider the smart home example which is subset of IoT. Smart home is localized in space, provide services in a household. Devices in the Smart Home are federated into a network and furnish means for entertainment, monitoring of appliances, controlling of house components and other services. In the scenario of trusted smart home service, data assets would include data stored on the end user device, data typed by the user, the data stored in database or data transmitted over communication medium (E.g. location data). Also passkey which authorizes owner to access home must be protected from unauthorized access and its integrity should be maintained as well as authentication needs to be taken care. These assets are expected to be the main targets of a malicious attack. Devices or users are granted access rights to protected resources and services. These rights are implemented as credentials which must be safeguarded by an attacker. Detail use case and misuse case for smart home system is described in Section 2.2.2.

2.2.2 Use Cases and Misuse Cases

The actor in use case and misuse case in the scenario of smart home includes: Infrastructure owner (smart home), IoT entity (smartphone device or software agent), attacker (misuser) and intruder (exploiter).

- Access control

This operations deal with issuing access rights to protected resources and systems. Granting of voting credentials, passkey issuance and granting of access rights are few examples.

In Table 2.1, use case and misuse case clearly depicts how the smart home is prone to attack for access control operations. There are several use cases possible for different scenario cases. In the sequel, different threats collected and control objectives are summarized below:

a) Access rights granted to unauthorized entity

Access rights may be granted to an unauthorized actor if an attacker is able to subvert the access control process. One way to do this may be done through impersonation, social engineering, by sending targeted e-mails requesting for access rights etc.

Table 2.1 Use case and misuse case for access control

Use Case		Misuse Case	
Granting Access		Access Rights Granted to Unauthorized Device	
Description	Actor gets access to resource	Description	Misuser granted access rights directly
Precondition	Actor has access privilege	Precondition	Actor has sufficient privilege to perform this operation
Success flow	Actor confirms identity of requesting actor Credential verification Granting of access	Assumption	Misuser is able to impersonate a legitimate access requesting entity
Actor	Infrastructure owner/ requesting device	Actor	Misuser
		Assets	Access credentials

1. Access rights should only be granted to actors after verification of their identity.
2. Provision of filters or other equivalent mechanism should be installed to identify type of actors.
3. If no formal verification of identity possible, then there should be alert provision before granting access rights.

b) Corruption of access credentials

Depending on the chosen solution used for representing access right credentials, attacker is able to get hold of certain options. If the credentials are stored with the device they may be subject to manipulation by a malicious entity (user / device). This can be used to gain extra privileges by tampering with the credential's data structure.

1. A secure design should be used to implement credential storage. Credentials should be stored on a device or should be generated depending on the context, to avoid tampering by an attacker.
2. Otherwise integrity of credentials should be protected by cryptographic means.

c) Unauthorized data transmission

Unauthorized data sent by an entity of an IoT network may lead to a breach of privacy. Even the number or the different types of devices constitute private data, measures to be followed are as follows:

1. Traffic monitoring should be detected
2. Integrity of messages should be taken care

d) Denial of service (DoS) attack

If a successful DoS attack can be mounted against the smart door software agent or then notification alerts about the door open status can be suppressed. If this attack is combined with the first one then access to the Smart Home can be obtained.

1. Software agent should be proofed against tampering and DoS attacks.

e) Man-in-the-middle attack

Federation over insecure network may lead to eavesdropping which may be exploited further for data theft or identity theft.

1. Federation requests should only be accepted from entities after verification of their identity.
2. Strong encryption techniques should be employed to protect confidentiality of identity or location to ensure identity/location privacy.

A threat analysis presented may also comprise a risk analysis where severity and probability can be estimated and then risk can calculated for each threat. The objective of this use case and misuse case-based threat modeling is to incorporate them in the security assessment of IoT networks.

2.2.3 Activity Modeling and Threats

The activity modeling of IoT attacks is use to understand the sequence of actions taking place when the attacks are happening. When there is solution for authentication and IdM, its needs to be analyzed for adversary models. Adversaries have been defined in many ways [10, 11] in literature. If we know, and understand possible attacks, we can decide countermeasures, and mitigation to deal with those attacks [12]. Security threats are designed using attack tree where root node represents attack goal, leaf nodes represents different ways of achieving the goals, and internal nodes represents attack steps. Discovery and avoidance of threats, and attacks in the system or networks is the most important task. To this purpose, we can use attack modeling like a graph-based collaborative attack modeling [13] in which sample of attack scenarios are used to demonstrate the attack steps.

Privacy model is required for privacy protection against adversaries. Adversary is someone whose purpose is opposed to, or conflict with the system functionality. Adversary is classified based on their capabilities like nature as active, or passive, static, or adaptive, computational ability, mobility, and byzantine. Adversary models are subject to change depending on the underline application [14]. Adversaries are classified based on their capacities into three types as [14]:

1. *Weak passive:* These are passive eavesdropper with limited capacity, and cannot gain whole control over transmission path.
2. *Strong passive:* These are passive eavesdropper, and can gain whole control over transmission path.
3. *Strong active:* These are active eavesdropper with the ability of compromising intermediate source, and destination.

2.2 Threat Modeling   41

In the view of these adversaries, as shown in Figure 2.4, IoT is prone to man-in-the-middle attack, impersonation which can cause DoS attack, and replay attack. In IoT, any device can communicate with any other device through wireless media, or through Internet. Possible communications are between device-to-device, human-to-device, human-to-human giving connection between heterogeneous entities, or network. Figure 2.4 presents general use case of IoT where MobileEntity(x): A mobile device represents an entity i.e. any device in the network which communicates with other entities of the same type, or of different types via Internet, or direct. Mobile Entity 1, 2, 3 represent three different and most probable scenarios in the system

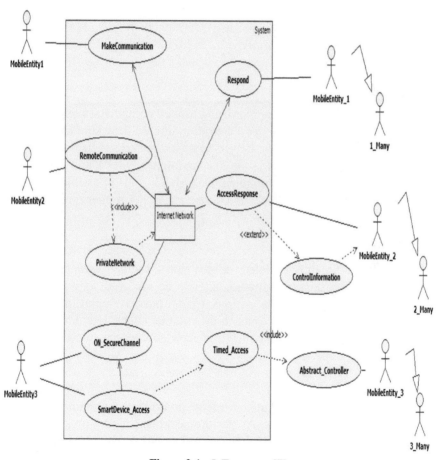

Figure 2.4   IoT use case [7]

of communication. Different possible attacks in IoT communications are described below.

- **Man-in-the-middle attack**

When the devices are commissioned into a network, keying material, security, and domain parameters could be eavesdropped. Keying material can reveal secret key between devices and authenticity of the communication channel could be compromised. Man-in-the-middle attack is one type of eavesdropping possible in commissioning phase of devices to IoT. Key establishment protocol is vulnerable to man-in-the-middle attack and compromise device authentication as devices usually do not have prior knowledge about each other. As device authentication involves exchange of device identities, identity theft is possible due to man-in-the-middle attack. A sample of man-in-the-middle attack is shown in Figure 2.5.

- **DoS attack**

All the devices in IoT have low memory, and limited computation resources, thus they are vulnerable to resource enervation attack. Attackers can send messages, or requests to a specific device so as to consume their resources. This attack is more daunting in IoT as the attacker might be single, and resource constrained devices are large in numbers. DoS attack is also possible due to man-in-the-middle attack. A sample of DoS attack in IoT scenario is shown in Figure 2.5.

- **Replay attack**

While exchange of identity related information or other credentials in IoT, this information can be spoofed, altered or replayed to repel network traffic. This causes a very serious replay attack. Replay attack is essentially one form of active man-in-the-middle attack. Replay attack can be prevented by maintaining the freshness of random numbers, for example by using time stamp or nonce by including Message Authentication Code (MAC) as well. Sample of replay attack is shown in Figure 2.5.

To this purpose, authentication, and access control are the main security issues which are to be addressed. As per the adversary model presented, a strong active type of adversary which is most powerful needs to compromise the proposed authentication scheme.

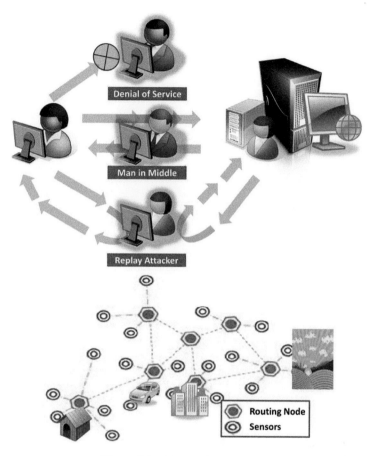

Figure 2.5 IoT attacks scenario [7]

2.2.4 IoT Security Tomography

As presented in [15], this section presents security tomography in the IoT context. It is classified according to attacks addressing to different layers. Here we have considered layers namely transport layer, network layer, MAC layer and RF layer. Figure 2.6 illustrates the IoT security tomography and the layered attacker model.

I) Threats on transport layer
Depending on types of protocol used in transport layer attacks are classified. By interfering connection of connection based protocol (E.g. TCP) can be

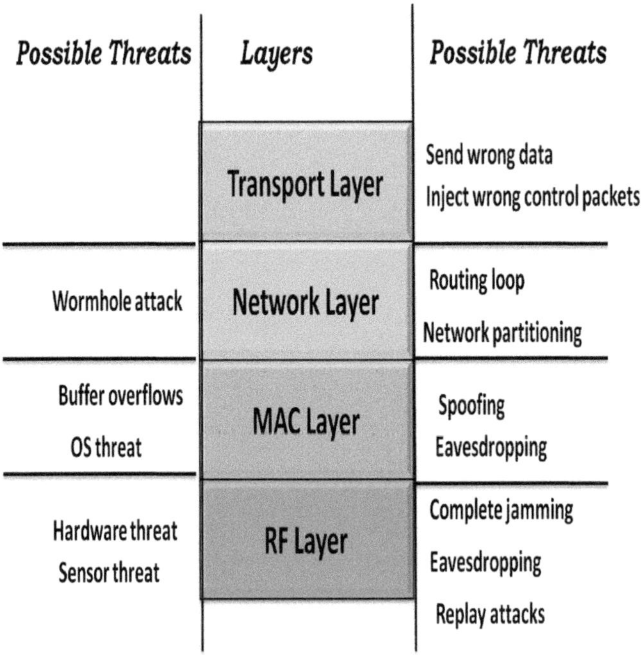

Possible Threats	Layers	Possible Threats
	Transport Layer	Send wrong data Inject wrong control packets
Wormhole attack	Network Layer	Routing loop Network partitioning
Buffer overflows OS threat	MAC Layer	Spoofing Eavesdropping
Hardware threat Sensor threat	RF Layer	Complete jamming Eavesdropping Replay attacks

Figure 2.6 IoT security tomography and layered attacker model

tampered. Connectionless protocols like UDP usually do not have sequence or session number and also do not provide error and flow control. That is why packets are getting in incorrect order or some packets can also be lost. Threats possible on transport layer are as follows:

1. **Send wrong data:** Sending of packets with wrong information enforces error correction on the receiver node, which requires CPU power and therefore results into additional energy consumption.
2. **Inject wrong control packets:** Injection of packets into a running connection can enforce to close, de-sync or interfere the connection [15]. The packet injection process allows an unknown third party to disrupt or intercept packets from the consenting parties that are communicating, which can lead to degradation or blockage of users' ability to utilize certain network services or protocols [16].

II) Threats on network layer

Attacker in the network layer intends at disturbing and degrading the routing service. Threats on network layer are as follows:

1. **Routing loop:** The attacker simply forges new routing packets or modifies existing ones to create cyclic packet trajectory. These packets cyclically traverse nodes and they never reach their destination. This type of attack shortens network lifetime by consuming valuable network resources [15].
2. **Network partitioning:** This type of attack has the most serious impact on the overall routing service. The adversary separates the network to disjoint set of nodes that cannot reach each other. It can be achieved by injecting falsified routing packets or simply by other interventions that cause some nodes, which are cut-set nodes the removal of which cause the network to partitioned, to be break down sooner because of energy depletion [15].
3. **Wormhole:** The wormhole attack may be launched by a single or a pair of collaborating nodes. In commonly found two ended wormhole, one end overhears the packets and forwards them through the tunnel to the other end, where the packets are replayed to local area. It either drops or selectively forwards the packets, leading to network disruption. [17].

    A wormhole attack is caused by one, two or more number of nodes. In one ended wormhole, fake neighbors are created by establishing high power node. Two ended wormhole is the most commonly found type of wormhole. It can be set via an out-of-band channel between two nodes or by encapsulating packet. In the latter case the nodes on way from first wormhole end to other, cannot increment hop count and thus wormhole attack becomes the result [18].

III) Threats on MAC layer

Attacks to the medium access layer are based on the shared medium characteristic of the broadcast medium. More precisely, the focus on security threats of this category arises from point-to-point local broadcast over a shared medium. Possible threats on MAC layer are as follows:

1. **Spoofing:** Unauthorized parties may participate in the network, e.g. by spoofing their identities and using the identity of another device.
2. **Eavesdropping:** Eavesdropping by nature is possible on a broadcast medium. An attacker who is physically located within the transmission range of the sender device receives all the traffic and can read it unless there is no protection mechanism in place

IV) Threats on RF layer

Attacks on the radio can be subdivided into jamming, spying and replay.

1. **Complete jamming:** Due to increasing level of noise level, the communication becomes difficult and eventually impossible anymore. For example the communication of a group of sensors can be disabled in order to avoid propagation of alarm triggering information [15]
2. **Eavesdropping:** Broadcasted RF-information can be received. It is an initial step of gathering transmitted information.
3. **Replay attacks:** Recorded RF-information can be replayed at wrong time even without knowing the content of the packets. It confuses the receiver or can set it into wrong state.

Different threats like sensor threat, hardware threats and OS threats are also possible in IoT.

I) OS threat: Attacks on the OS can be categorized in attacks that want to change the behavior of the nodes and in attacks that want to disable the node or a critical service of the OS. Attacks that change the behavior of the node can be categorized into attacks that can be done remotely and attacks that require direct contact to the hardware [15].

1. **Buffer overflows:** Such attacks intentionally force a buffer to store more data than it is intended to hold. The overflowed data can alter binary code and therefore the behavior of the node.
2. **Direct reprogramming:** Sensor nodes that can be directly accessed (collected) can be manually reprogrammed by changing applications, OS, OS services that are stored in RAM or ROM.

II) Hardware threat: Here attacks are classified by direct electrical access to the internal components of the device, e.g.: micro-probing.

III) Sensor threat: Such attacks are classified by a direct electrical access to the internal components of the device, for example by micro-probing. This attack is also called as falsified sensor reading [15].

2.3 Key Elements

Threats are potential causes of an event that could breach security and causes possible harms. Some time there is situation where protection mechanism becomes subject to harm. To avoid these, some security policies are decided. Threats are occur due to weaknesses in the mechanism which implementing a particular security policy.

Security in IoT environment should address the following main issues [19].

- Enabling smart and intelligent behavior of networked objects.
- Preservation of privacy for heterogeneous sets of objects.
- Decentralized authentication and trust model.
- Energy efficient security solutions.
- Proper authentication of the objects within the network.
- Security and trust for cloud computing services.
- Data ownership.

Key elements of security are,

1. Authentication (Identity establishment) which can set up proof of identities;
2. Access Control specifies who can access;
3. Data and message security are referred as data integrity and confidentiality;
4. Prevention from denial of taking part in a transaction, whether as an initiating or a receiving party, known as non-repudiation [20] and availability states that resources or information should be accessible to authoritative party at all instant.

Further details of the elements of system security are explained in following points.

2.3.1 Identity Establishment

As mentioned above, secure entity identification is known as identity establishment which is also referred as authentication. Authentication is an identity establishment between communicating parties (devices) or entities. Entity can be a single user, a set of users, an entire organization or some networking device. Identity establishment is ensuring that the origin of an electronic document and message is correctly identified. Identities are the windows through which users interact with their devices, and consume services in today's world. Before any service is delivered, it is customary to verify a digital identity of the user requesting that service (user identity) and also the identity of the entity offering the service (service identity). In IoT world, this concept of identity extends to things. Ensuring that things have a means to be identified is critical to assure users that their interactions with things are safe.

Many security mechanisms have been proposed based on private key cryptographic primitives due to fast computation and energy efficiency.

Scalability problem and memory requirement to store keys makes it inefficient for heterogeneous devices in IoT. A public key cryptography based solution overcomes these challenges because of its high scalability, low memory requirements and no requirement of key pre-distribution infrastructure [7].

In [21], the author presented ECC based mutual authentication protocol for IoT using hash functions. Mutual authentication is achieved between terminal node and platform using secret key cryptosystem introducing the problem of key management and storage. Self-certified keys cryptosystem based distributed user authentication scheme for WSN is presented in [22], where only user nodes are authenticated

2.3.2 Access Control

Access control is also known as access authorization or simply authorization. The principles of access control determine who should be able to access what [2]. Access control prevents unauthorized use of resources. To achieve access control, entity which trying to gain access must be authenticated first. According to authentication, access rights can be modified to the individual. Introducing a new device, or user, and achieving authentication and access control to devices resources in IoT is critical. As IoT is ad-hoc, and dynamic in nature, efficient, and a robust mechanism of secure access to resources must be deployed with distributed nature. Traditionally, access control is represented by an Access Control Matrix (ACM), in which the column of ACM is basically a list of objects, or resources to be accessed and the row is a list of subject or whoever wants to access the resource. From this ACM, two traditional access control models exist, i.e. Access Control List (ACL) and capability-based access control. Due to unbound number of devices, and services, scalability, and manageability issues are daunting in IoT. Various access control models and their applicability in the context of IoT is presented and discuss in detail in chapter 6 of this book.

2.3.3 Data and Message Security

Data security is mostly concerned with source authenticity, modification detection, and confidentiality of data that is being processed in-memory, or while residing on a storage medium or during transmission over a computer network. Combination of modification and confidentiality of message is not enough for data integrity, but origin of authenticity is also important. Location privacy is equally important risk in IoT. To ensure location privacy, communication and reference signal integrity needs to be maintained.

Communication confidentiality and privacy of localization and tracking data is highly sensitive in IoT amalgam. There should not be any way for an attacker to reveal identity or location information of device to ensure privacy. When the contents of message are changed after sending this message from source but before reaches at destination then we can say that integrity of message is lost. Integrity services assure that data sent are received as no duplication, insertion, modification, or replays.

2.3.4 Non-repudiation and Availability

Non-repudiation (NR) is one of the security services (or dimensions as defined in the document X.805 by the ITU) for point-to-point communications [23]. Non-repudiation of action is the process by which an entity (sender or receiver) is prevented from denying a transmitted message. So when message is sent, receiver can prove that initiating sender only sent that message. Similarly sender can prove that receiver got the message. To repudiate means to deny. For many years, authorities have required to make repudiation impossible in some situations. You might send registered mail, for example, so the recipient cannot deny that a letter was delivered. Similarly, a legal document typically requires witnesses to signing so that the person who signs cannot deny having done so [24].

Both X.800 and RFC 2828 define availability to be the property of a system or a system resource being accessible and usable upon demand by an authorized system entity, according to performance specifications for the system i.e., a system is available if it provides services according to the system design whenever users request them. For example banking customers should be able to check their balance at any time so server must be available at all time.

Availability is ensured by strictly maintaining all hardware, repairing immediately whenever require. It also prevents the bottleneck occurrence by keeping emergence backup power systems and guarding against malicious actions like Denial of Service (DoS) attack.

2.3.5 Security Model for IoT

Integrated and interrelated perspective on security, trust, privacy can potentially deliver an input to address protection issues in the IoT [6]. Therefore cube structure is chosen as a modeling mechanism for security, trust and privacy. A cube has three dimensions with the ability to clearly show the intersection thereof. Therefore a cube is an ideal modeling structure for depicting the convergence of security, trust and privacy for the IoT. In IoT access information, required to grant/reject access requests, is not only complex but also composite in nature. This is a direct result of the high level of interconnectedness between things, services and people. It is clear that the type and structure of information required to grant/reject such an access request is complex and should address the following IoT issues: security (authorization), trust (reputation), privacy (respondent). This is depicted in Figure 2.7.

Current Internet security protocols rely on a well-known and widely trusted suite of cryptographic algorithms: the Advanced Encryption Standard (AES) block cipher for confidentiality; the Rivest-Shamir-Adelman (RSA) asymmetric algorithm for digital signatures and key transport; the Diffie-Hellman (DH) asymmetric key agreement algorithm; and the SHA-1 and SHA-256 secure hash algorithms. This suite of algorithms is supplemented by

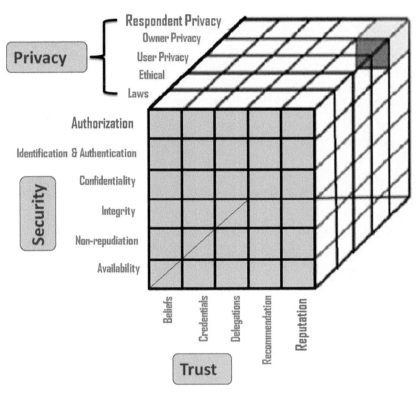

Figure 2.7 Security model for IoT

a set of emerging asymmetric algorithms, known as Elliptic Curve Cryptography (ECC). Adoption of the ECC algorithms has been slowed by significant IPR concerns, but publication of RFC 6090 and recent IPR disclosures may encourage adoption [25].

2.4 Conclusions

In the environment of IoT interactions between devices, user and service provider should be secure, in spite of the type of devices used to access a service. We must assure that enough privacy and security is available before the technology gets deployed and becomes a part of our daily life. A lightweight, distributed and attack resistant solution are the most vital properties for the security solution in IoT. This puts resilient challenges for IdM and access control of devices. The access control is very important for successful realization of IoT, especially due to the dynamic network topology, and distributed nature.

The incremental deployment of the technologies that will make up the IoT must not fail what the Internet has failed to do: provide adequate security and privacy mechanisms from the start. We must be sure that adequate security and privacy is available before the technology gets deployed and becomes part of our daily live. Security requirement and threat taxonomy insist to go for trusted platform module which offers facilities for the secure generation of cryptographic keys. Threat modeling, threat analysis and activity modeling of threats presented in this chapter are used to understand the sequence of different possible attacks, and accordingly it is easy to consider what action has to takes place when the attacks are happening. Identity establishment, access control, data and message security, non-repudiation and availability are the most vital elements which needs to be considered for the security in the IoT. This chapter also presents security model for the IoT with the convergence of trust and privacy.

Referneces

[1] T. Heer, O. Gracia-Morchon, R. Hummen, S.L. Koch, S. Kumar, and K. Wehrle, "Security challenges in the IP-based Internet of Things," In wireless Personal Communications, 61(3): 527–542, 2011.
[2] Atul Kahate "Cryptography and Network Security", Tata McGraw-Hill
[3] http://www.webopedia.com/TERM/D/DDoS_attack.html

[4] http://siliconangle.com/blog/2013/08/26/5-notorious-ddos-attacks-in-2013-big-problem-for-the-internet-of-things/

[5] siliconangle.com/blog/2014/01/23/the-internet-of-things-is-riddled-with-vulnerabilities/

[6] Sachin Babar, Parikshit N. Mahalle, Antonietta Stango, Neeli R Prasad, and Ramjee Prasad, "Proposed Security Model and Threat Taxonomy for the Internet of Things (IoT)," In proceedings of 3rd International Conference CNSA 2010, Book titled: Recent Trends in Network Security and Applications - Communications in Computer and Information Science 2010 Springer Berlin Heidelberg, pp: 420–429 Volume: 89, Chennai – India, July 23–25 2010

[7] Parikshit N. Mahalle, Bayu Anggorojati, Neeli R. Prasad and Ramjee Prasad, "Identity Authentication and Capability-based Access (IACAC) Control for the Internet of Things," In Journal of Cyber Security and Mobility", River Publishers, Volume: 1, Issue: 4, pp: 309–348, March 2013

[8] C. Ding, L. J. Yang, and M. Wu, "Security architecture and key technologies for IoT/CPS", ZTE Technology Journal, vol. 17, no. 1, Feb. 2011.

[9] Hui Suo; Jiafu Wan; Caifeng Zou; Jianqi Liu, "Security in the Internet of Things: A Review," Computer Science and Electronics Engineering (ICCSEE), 2012 International Conference on , vol.3, no., pp.648, 651, 23–25 March 2012

[10] B. Wood, "An Insider Threat Model for Adversary Simulation," In Procedings of 2nd Workshop on Research with Security Vulnerability Databases, SRI Internaqtional, Santa Monica - CA, 20002.

[11] Paul Syverson, Gene Tsudik, Michael Reed and Carl Landwehr, "Towards an Analysis of Onion Routing Security," In Workshop on Design Issues in Anonymity and Unobservability, Volume 2009 of Lecture Notes in Computer Science, pp: 96–11, July 4 2001.

[12] B. Schneier, "Attack Trees," In Dr. Dobb's Journal., Volume :24, Issue: 12, pp: 21–29, 1999.

[13] J. Steffan, and M. Schumacher, "Collaborative Attack Modeling," In Procedings of 17th ACM Symposiyum on Applied Computing (SAC 2002), ACM Press, pp: 253–259, Madrid – Spain , March 10–14 2002.

[14] Parikshit N. Mahalle, Neeli R. Prasad, and Ramjee Prasad, "Object Classification based Context Management for Identity Management in Internet of Things", In International Journal of Computer

Applications, Volume: 63, Issue : 12, pp: 1–6, February 2013, Published by Foundation of Computer Science, New York, USA.
[15] Casaca, INOV, Dirk Westhoff, NEC Europe Ltd., "Scenario Definition and Initial Threat Analysis", Project name: Ubiquitous Sensing and Security in the European Homeland, 2006.
[16] http://en.wikipedia.org/wiki/Packet_injection
[17] Devesh Jinwala, "Ubiquitous Computing: Wireless Sensor Network Deployment, Models, Security, Tbreats and Challenges", in National conference NCIIRP-2006, SRMIST, pp. 1–8, April 2006.
[18] Buch, Dhara; Jinwala, Devesh, "Detection of Wormhole attacks in Wireless Sensor Network," Advances in Recent Technologies in Communication and Computing (ARTCom 2011), 3rd International Conference on , vol. 7, no. 14, pp. 14–15 Nov. 2011.
[19] Bhattasali, Tapalina, Rituparna Chaki, and Nabendu Chaki. "Study of Security Issues in Pervasive Environment of Next Generation Internet of Things." Computer Information Systems and Industrial Management. Springer Berlin Heidelberg, 2013. 206–217.
[20] Messaoud Benantar, "Access Control Systems: Security, Identity Management and Trust Models", Springer.
[21] Guanglei Zhao, Xianping Si, Jingcheng Wang, Xiao Long, and Ting Hu. A novel mutual authentication scheme for Internet of Things. In Proceedings of 2011 IEEE International Conference on Modelling, Identification and Control (ICMIC), pp. 563–566, 26–29 June 2011.
[22] C. Jiang, B. Li, and H. Xu. An efficient scheme for user authentication in wireless sensor networks. In 21st International Conference on Advanced Information Networking and Applications Workshops, pp. 438–442, 2007.
[23] https://www.nics.uma.es/research/non-repudiation
[24] http://searchsecurity.techtarget.com/definition/nonrepudiation
[25] Tim Polk, Sean Turner, "Security Challenges For the Internet Of Things", IAB, IETF Security Area Directors, February 14, 2011.

3
Identity Management Models

3.1 Introduction

There is a profound change today in the wireless communication with the increase in mobility of portable, yet powerful wireless devices capable of communicating via several different kinds of wireless radio networks. The requirement for identity is not adequately met in the networks, especially, given the emergence of ubiquitous computing devices that are mobile, and use wireless communications. IdM solution requires changes in the identifier format, and addressing mechanism. For IoT, it is envisioned that an incredibly high number of inexpensive pervasive devices surround us. Connecting all these devices to the Internet will involve the integration of multiple connectivity options based on the many designs, and deployment constraints. As discussed in the Chapter 1, the major factors are resource constrained devices with low energy, low bandwidth, low computational power, and distributed nature of IoT networks.

3.1.1 Identity Management

In IoT networks, normal things, or devices are a part of the whole network in order to collaborate, understand, and react accordingly as per the need. There are some objects, or things which get destroyed after some time, and therefore they do not require global unique identification. On the other hand, there are many types of objects like mobile devices or items in the mall which require unique identification. Meaning of an identity and design of an identifier in IoT context is one of the main issues in the view of resource constraints like energy, lifetime, end-to-end delay, memory, and routing overhead.

An identity is something which makes the thing distinguishable and delineate. Things under consideration only have one identity, but might be associated with many identifiers. These identifiers are used to distinguish between two things as unique entities and are context dependent as well. IdM in IoT include identifying things, assigning identifiers to them, performing authentication and managing access control.

3.1.1.1 Identifiers in IoT

An identifier discerns different users, places, or things within the context of specific namespace. The namespace plays an important role in defining an identifier because identifiers are always local to the current namespace. For example, user, and sensor both have identifiers. The user may be associated with a bank, an office, or home. Here the bank, office, and home are different namespaces, and each will have a different identifier. Each identifier is meaningful in the namespace, and only when associated with things being identified. Example for CAR entity and its identifiers are shown in below.

$$CAR = \{VIN, LICENCE\ PLATE, TYPE\}$$

CAR has three identifiers, and association of CAR with one of the identifiers is used depending on the context, and the namespace. Precisely, identifier can be defined in a generic way as having three parameters as

$$Identifier = \{Thing, Identifier, Namespace\}$$

e.g. {CAR, VIN, RTO_DB}, {SENSOR, NODE_ID, HOME_GATEWAY}, {TAG, EPCID, LOCAL_DB}

Things will be associated with many identifiers, and is shown in the Figure 3.1.

An attribute is a dedicated characteristic associated with an entity like sensor node, or object with RFID tag in IoT. As attributes are only going to be exchanged for association with an identifier, meaningful attributes of things need to be defined for IoT along with the scope rules. The attributes will vary from personal space to public space. Broadly, there are two types of attributes: persistent attribute which are permanent attributes of devices and non-persistent attributes which are temporary attributes of devices. We propose that each device should be associated with at least one persistent, and one non-persistent attribute, as both types of attributes will have different meanings in the local context.

3.1.1.2 Identification and identifier format

The association of identifiers with devices presenting an attribute is called as identification. For example, device is PDA with ID_1. This example includes accepting the association between device PDA, and its attribute as ID_1.

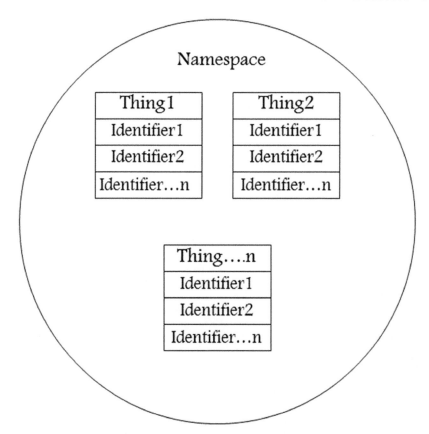

Figure 3.1 Things and identifiers in IoT

As discussed in the above section, things can have many identifiers, and each identifier has to be associated with it depending on the context. Identification is applicable to both devices and users, and it requires identifier. Devices are always acquiring some attributes, and authentication is referred as collection of proofs for attribute. When devices communicate with each other, or provide any service, they always provide some attributes along with the identifier to authenticate. Identification is represented as

$$\{\text{Thing identified, thing}\} \; \varepsilon \; \textbf{Namespace}$$

IdM is a set of processes that consist of identity binding, identity mapping and authentication. It involves management and exchange of device identity information, also known as digital identity. Precisely, we define IdM as management of identity followed by identity authentication, and attribute authentication. In IoT, each end user, service, or thing will be represented by an identity, and identity is a set of temporary or permanent attribute of devices. Depending on the context in use, the separate Context Identity (CID) can be used. In order to support context-awareness and applying namespace dependent identifier to device, utilization of context information is an important aspect. A General definition of context is, any information that can be used to classify the situation of an entity. An entity is a person, place, or object that is considered relevant to the interaction between a user, and an application. It is clear that such information is very important to select and apply appropriate identifier to device.

Figure 3.2 shows identifier format for devices in IoT as presented in [1]. Nomadic devices in IoT can join to public or private IoT. In this regard, it is essential to assign ownership to these devices. As devices can be people or information, and this classification must be present as one of the parameters in the proposed format. It should be easy to know that thing is RFID tag, sensor node, sensor network or PDA. For unique identification purpose, unique identifiers like EUI-64 bit of 802.15.4, EPC code [2], or any other unique identifiers are associated with this format. This format for devices should

ORI = <OBJECT 0>, <RESOURCE-1>|<OBJECT TYPE>|< GLOBAL NAMESPACE> | < LOCAL NAMESPACE > | <UID> | <CID>

Where:

<OBJECT 0>, <RESOURCE-1> = Indicates object is Thing or Service

<OBJECT TYPE> = Type of Object e.g. TAG | SENSOR | PDA

<GLOBAL NAMESPACE> = Indicates global Ownership / Interface

<LOCAL NAMESPACE> = Indicates local Ownership / Interface

<UID> = Unique identification number of device e.g. EUI – 64 of 802.15.4 | EPC code | UUID

<CID> = Context Identity

Figure 3.2   Identifier format for things

have association with the different attributes, and these attributes are decided on the namespace in which devices are being used. ORI represents object, or resource identifier. Object type field is used to differentiate between the types of object it is representing. This field essentially linked to CID field of identifier format. GLOBAL NAMESPACE field is used to indicate global ownership, or interface, and is very useful in mobility of the device, or thing. The significance of the LOCAL NAMASPACE field is to decide current status of the device. The UID represents the unique identifier for the object, or devices under consideration.

3.1.2 Identity Portrayal

IdM is a combination of processes, and technologies to manage, and secure access to the information, and resources while also protecting things' profiles. Identity of devices has been considered in a number of different ways in various literatures. From literature it is evident that, at high levels there are three dependent sets of object identity domain in any possible scenario of IoT. These three domains are individual, social and technological [3]. As these three domains are inter-dependent viz, one, two, or all three domains are applicable to object identity in IoT. As device identity in individuals and social domains is well defined, and established, efforts are required to define, and formulate device identity in technological domain, and in turn for IoT. In the current era of web, and Internet computing, IdM is oriented towards identity of either device, or user, but in IoT, mapping between IoT device identity, and context identity is required. Devices under consideration have only one identity but might be associated with many identifiers. These identifiers are used to distinguish between two things as unique entities and are also context dependent.

As described in [4], an identity refers to the abstract entity that is identified. An identifier, on the other hand, refers to the concrete bit pattern that is used in the identification process. We define IdM as accomplishment of three phase's mentioned below concern with thing identity in IoT. Identity portrayal is done through the following phases:

- **Substance:** Identity is established i.e. authenticated through the identifier

In this phase, credentialisation, and associated process of credentialisation is considered. Credentialisation encompasses authentication, identification and assignment. Authentication is signalled by identifiers for identity establishment. Identification is typically signalled by its attributes.

60 Identity Management Models

- **Content:** Identification, and communication

This phase deals with how identity relates with communication. As identifiers are ubiquitous in IoT, there are numerous objects in the surrounding, validating association between object, and there is a need of group authentication schemes in the context of IoT.

- **Use:** Appropriate identity is used in various context of IoT

This phase explains how identity is expected to perform, and how other objects perform towards particular objects. Access control is taken care of in this phase of identity portrayal.
Aforementioned identity portrayal is depicted in the following Figure 3.3.

Persistent identifiers are required to establish identity between devices when communication is remote in time and space, else, non-persistent identifiers are sufficient. Unique identity of device can also be determined by data collected from various sources. The profile represents interest domains such as personal profile, private profile, and trust profile. In IoT, it is necessary to create a profile of identification attributes to describe devices. Building

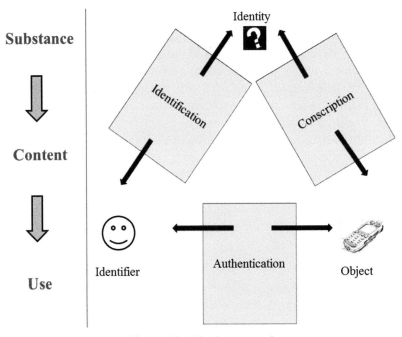

Figure 3.3 Identity portrayal

profile of things in nomadic IoT is expensive, hence there is a need of architectural approach for IdM. The main purpose of the identifier is to uniquely identify things, objects or devices. This is applicable to daily life as vehicles are uniquely identified by number plates and in digital world, network devices are uniquely identified by the MAC address. Identifiers are manageable representation of devices, and enable quick and reliable access to it. The uniqueness of these identifiers is based on the contexts or it is also possible to provide uniqueness with the help of additional ad-on attributes in identifiers. There are different ways to construct identifiers which are listed below from 1–5.

1. Using random data
2. Hierarchical identifiers
3. Encoded identifiers (E.g. Time stamp or other contextual information)
4. Cryptographic identifiers (E.g. Hash or digest)
5. Hybrid identifiers (Mix of few of these from 1–5)

As presented in the above section, network devices, or electronic objects are identified by various identifiers like RFID identifier, MAC address, and IP address, URL or URI, and refers to different layers of ISO / OSI model. In the Internet world, network devices are identified by IP addresses, and services are identified by URLs, but this approach only works for homogenous environment. In IoT, RFID tags do not have IP addresses, and therefore respective services cannot be accessed by URLs.

3.1.3 Related Works

Different identity schemes have been proposed in IoT and it is predicted that it is dubious to have common identification schemes globally [2]. Existing identification schemes in the context of IoT are listed below:

- RFID Object Identifier
- EPCglobal
- Short-OID
- Near Field Communications Forum
- Handle and ODI
- Ubiquitous Code
- URL as an identifier
- IP address as an Identifier

Limitations of these identification schemes are listed in the Table 3.1 give below [2].

Table 3.1 Limitations of different identification schemes [2]

RFID Object Identifier
• Lack of resolver system to address the different OID structures • Centralized in nature • No marketing budget for an ISO standard
EPCglobal
• Restricted to GS 1 domain only • Lack of multilateral security and confidentiality • At thing level, there are limited and uncertain data carrier options • Cost involved for few retailers using the system is more
Short-OID
• Lack of proper resolver system to address this OID structure • Lack of domain specific differentiation because common root could not enable this differentiation • Similar to RFID OID
Near Field Communications Forum
• Air Protocol specific • Data capture integration with other tags is low • Much similar to 2D bar codes
Handle and ODI
• Require additional infrastructure overload for additional application • Isolated from data carriers and not suitable for physical objects
Ubiquitous Code
• Weak due to reverse logic of the code declaring the data transfer • Not powerful as EPCglobal
URL as an Identifier
• Long in length, and not suitable for data capture • Lack of security
IP Address as an Identifier
• Not suitable to lightweight objects with resource constraints • Scalability problem

State of the art shows that there has been a lot of work for IdM, and identities, but none of the work addresses IoT. Things under consideration have only one identity but might be associated with many identifiers. These identifiers are used to distinguish between two things as unique entities and are also context dependent. Different identity schemes have been proposed in IoT, and it is predicted that it is dubious to have common identification schemes globally [4]. Identification schemes for RFID Object Identifier, EPCglobal, Short-OID, and Near Field Communications Forum have been studied in [2]. In [5, 6], the author addresses the IdM problem in IoT with challenges, and presents naming, and addressing as one of the main issues for IoT. Verifying device ownership and identity by digital shadowing is presented in [7] where the user presents his/her virtual identity onto logical nodes. Virtual identities are based on the notions that the user's device acts on his/her behalf but does not store his/her identity. Only virtual identity representing information is projected, but addressing and implementation details are left unaddressed. An author presents the domain trusted entity where each identity is managed by a trusted entity of its corresponding home domain that keeps it under the preferences set by its holder. This approach is not suitable for futuristic IoT due to its dynamic topology, and distributed nature. Use of clustering for efficient resource management in IoT is proposed in [8] achieving lifetime of network, scalability, and reduced packet delay. Multi-hop clustering protocol for WSN without addressing mobility is presented in [9]. There have been many attempts on the solution for hierarchical addressing but all the solutions are focusing on IP networks, and the Internet domain level in the current Internet, and not suitable for IoT [10–12]. The DNS is a hierarchical naming system built on a distributed database for computers, services, or any resource connected to the Internet, or a private network [13]. The DNS is not suited for critical infrastructure and is prone to spoofing and authentication problem. Meanwhile, the Distributed Hash Table (DHT) is adopted as the underlying structure to construct the basic UID management methodology [14]. The problems using DHT for IdM are achieving load balancing while mapping keys to nodes, and forwarding lookup for a key to the appropriate node.

Current IdM solutions are mainly concerned with identities that are used by end users, and services to identify themselves in the networked world (e.g. Liberty Alliance [15], OpenID [16], etc.). These solutions provide user attributes, and authentication as a service to relying parties. It is a complex, and dynamically developing area due to its importance in online communities. The main IdM solutions focuses on the definition of IdM life cycle, definition of service integration with identity providers, the establishment of SSO mechanisms to define identity federations, and exchange of authentication information, and attributes with respect to end users, and services. This principle is adopted by many of the existing solutions like Shibboleth [17], Liberty Alliance [15], OpenID [16], WS-* [18] etc.

The Internet players and the Telco industry have been developing their IdM solutions along different paths to address different needs. In the Internet, the focus is more on providing solutions for the end user to access services, while in the Telco world, it is more the case of identifiers, and authentication, since deciding which entity is allowed to connect to the network is very important here. With the convergence of the Internet, and Telco worlds, these paths are merging with each other more, and more. Examples of efforts in this direction are the solutions developed in standardization organizations like 3GPP (e.g. GAA [19]), or in European projects like FIDELITY [20], SPICE [21] (e.g. GBA-SAML) and SWIFT [22]. The addition of devices in this space require that the concepts developed so far have to be extended and improved to include the scenarios made possible in IoT.

3.2 Different Identity Management Models

It is essential to derive taxonomy of different identity models depending on the scope of an identity as well as local and the global context in which an identity is represented and used. The notion of IdM in the context of the IoT is analogous to daily life practices to large extent. IoT objects/devices can have direct knowledge of other objects/devices which can be identified. The scope and life time of these things and their identities varies from context to context. As in daily life practice, an individual can be known to only her/his family, or close friends/neighbors local workplace, another individual can be known to his/her locality or bigger scope and some individuals are known across the globe. As presented in the above section, identity of the IoT objects will be context-aware and can be known locally, across the ubiquitous network or globally. In the sequel, IoT objects/devices can also be associated with multiple digital identities (referred as virtual identities) [7] in the same way as an individual has account number as an identity in the context of bank, employee id in the context of workplace and consumer number at electricity office. There is a need that these devices should be uniquely and unambiguously identified in multi-context IoT. As shown in the identifier format (Figure 3.2), each object is uniquely identified by a set of attributes, referred as profile. Based on the scope and space of identity, identity taxonomy in the order of increasing scope is as follows and described below:

- Local identity
- Network identity
- Federated identity
- Global identity

3.2.1 Local Identity

In the centralized architecture like smart home or client server paradigm, identity is local in nature. In centralized computing, a host system maintains and manages local database of identities. In the envisaged IoT context, smart

3.2 Different Identity Management Models

home is an example of centralized computing where all the devices in smart home are registered in the local database and if external device or entity wishes to join the system, it is first required to acquire an identity from server and an entry is to be made in registry. The system also checks for the duplication of an identity being issued in order to maintain uniqueness. In centralized computing, addition and deletion of identities is simple and independent of other operations. Generally identities assigned are flat in nature and scalable in nature. Figure 3.4 depicts the high level view of local identity model. Figure shows that in local identity model, a central object registry is maintained for all the terminals or devices connected to it. As in case of smart home scenario, all the devices in one context are registered with one central database and can be shared across the system.

Advantages of local IdM models are listed below:

- **Simplicity:** As one central entity (software agent or admin) is responsible for issuing and registering identity, manageability become easy. Generally, flat addressing or naming mechanism is adopted in local identity model and identity establishment process depends on the credentials

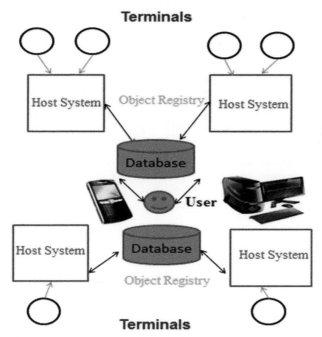

Figure 3.4 High level view of local identity management model

provided by objects. These credentials are then compared with the similar details stored in object registry/database. Local scope of identity and flat addressing makes it simple and manageable.
- **Manageability:** As the central entity is responsible for assigning an identity, object registry can assign and store the identities based on its capacity. But at the same time, with increasing number of objects in the scalable IoT networks, performance becomes the bottleneck.
- **Flat addressing:** As the scope of an identity in local IdM is local, flat addressing is useful for such system but it result into name collision.

In spite of these advantages, local IdM model have few drawbacks. Decisions and action in the context of local identity are more time consuming. There is also increased dependency and vulnerabilities. Greater number of objects that rely on one central registry causes the problem of single point of failure. In this type of model, there is more delay in the response time towards identity assignment and establishment as one central system is responsible for it. This makes the identity provisioning slow. Secrecy is hard to maintain in local identity models as all identities are stored at one central location.

3.2.2 Network Identity

In distributed IoT networks, the classical and centralized mechanism does not suffice because of the distributed nature of the device-to-device communication. IdM is one of the main issues in IoT because such networks could be both distributed, and dynamic in nature. In IoT, each device will have to assume that arbitrary devices can establish direct, ad-hoc communication with it. This distributed nature of IoT lead to the concept of network identity. In network identity model, identity is authenticated to the network of devices rather than an individual host or entity. Once the identity establishment is done to the network as explained, this identity will flow across the network providing services or resources without explicit identity establishment again. Thus, identity of an individual object will remain same in the participating network.

Network identity models also have an ability to establish cross-network identity or in most of the cases network identity is confined to single domain. Figure 3.5 represents high level of network identity model. In IoT network A, identity is confined to one domain, while in the IoT network B, identity is used across two IoT sub-networks.

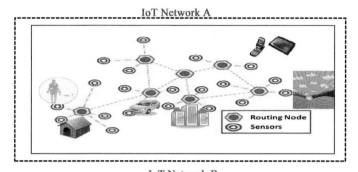
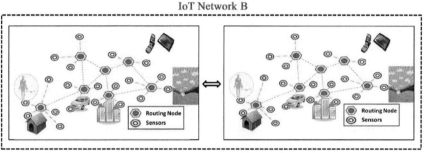

Figure 3.5 High level of network identity model

3.2.3 Federated Identity

Identity "Federation" is known term within the web security world and refers to management of a network/web object's identity across different domains. The main reason of enabling federation in the network/web environment is that the work flow of the system often requires an object for which identity is established in one domain to be established in other domains as well. First of all, the identity in the web-based system refers to a user's identity while in the IoT; identity refers to a device or "thing". Therefore, the interaction of identities in IoT is in the form of device-to-device communication [23].

An example of federated IoT network is depicted in Figure 3.6. Three IoT network domains are considered, i.e. private user, retail shop, and goods producer network, and two IoT-Federated networks are considered. An IoT network domain consists of one or more IoT cluster and the inter-cluster communication can be done either through the Internet infrastructure as shown in Figure 3.6 or through a wireless ad-hoc connection. Device-to-device communication within a cluster, i.e. intra-cluster communication, can be carried out by using different wireless access technology, e.g. RFID, ZigBee, Bluetooth, Wi-Fi, etc.

68   *Identity Management Models*

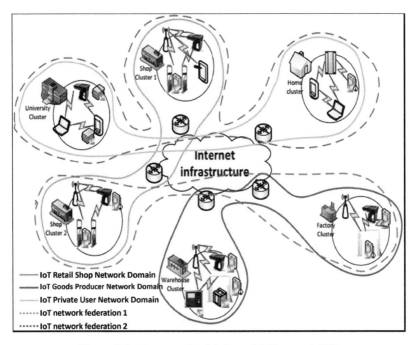

Figure 3.6 An example of federated IoT network [23]

There are different ways of accomplishing federated identity. In the federated networks, devices undergo single registration process. If the registration is performed more than once, then redundancy of the profiles is to be avoided. There are different federation topologies and they vary in the way profiles are created and managed. Federation topologies are as follows:

- *Local profiling:* All devices are registered with the IdM infrastructure of the local networks. Profiles of these devices are entirely managed by local network and local identity model is used for local profiling. Example of local profiling is smart home scenario.
- *Distributed profiling:* In this scheme, devices complete the registration process with the home IdM infrastructure and when needed, new profiles of the same devices can be created in new network. These profiles will be specific to the new network due to the need of new attributes. In the sequel, the profiles become distributed across multiple networks and attributes synchronization need to be taken care.
- *Third party profiling:* In this scheme, the trusted third party within established federation is involved for creating and managing profiles.

This alleviates the load from member networks from the registration. Trust management is an important issue to be taken care in third party profiling. The advantage of this scheme is that it is scalable in nature as more and more IoT networks can connected to trusted third party.

3.2.4 Global Web Identity

With the emergence of World Wide Web (WWW) [25] and the popularity of online social networks [26], global identity is a need of today. Web identity is uniquely identified throughout WWW and it is identifiable via Universal Resource Identifier (URI) [27]. Due to increasing number of users on WWW and online social network, it is important to keep unique identity of users as well as services. In the context of IoT networks, web identity is ubiquitous in nature and web identity information should be capable of uniquely resolving to one entity. Also it should be easy to maintain local and global context across various IoT networks. Due to evolution of the web from global library to global marketplace, secure and privacy-aware global IdM services are required.

3.3 Identity Management in Internet of Things

In such a world, the greater scale and scope of IoT increases the options in which a user can interact with the things in his/her physical and virtual environment. This broader scope of interactions enhances the need to extend current IdM models to include how users interact with devices as well as devices interact with other devices. Users interact with their devices and consume services in IoT through verified identity. In IoT, this concept of identity extends to devices/things. Compared to today's world, where interactions with devices and services are restricted by ownership and subscription, IoT users are able to discover and use devices that are public, add things temporarily to their personal space, share their devices with others, and devices that are public can be part of the personal space of multiple users at the same time. Secure interaction in and with IoT, secure data management and exchange, authentication, distributed access control and IdM of the devices are the main challenges.

In the context of IoT, IdM refers to the process of representing, and reorganizing entities, authentication, and access control. Requirement for identity is not adequately met in networks, especially given the emergence of ubiquitous computing devices that are mobile, and use wireless communications. IdM solution requires changes to the identities, and identifier formats for addressing. As computing technology becomes more tightly coupled into dynamic and mobile IoT, security mechanism becomes more stringent, less flexible, and intrusive. Scalability issue in IoT makes IdM of ubiquitous things more challenging. In IoT, each real thing becomes virtual which means that each entity has locatable, addressable and readable foil on the Internet. We need to identify resources, devices, agents, relationships, mappings, properties, and namespaces and provide identity securely. Traditional security solutions are not suitable due to resource constraints and scale to IoT's amalgam of context and devices.

It should be noted that most research has focused on IdM issues in the Internet and web computing era by only orienting users. Current IdM solutions are designed with the expectation that significant resources would be available and applicability of these solutions to IoT is unclear. Even the fundamental question of how well the IdM problem in the resource constrained IoT would be solved conceptually has been given little attention. A more insight on architectures and challenges for IdM is given in [28–30]. There are different IdM models which are discussed in the next section.

3.3.1 User-centric Identity Management

In user-centric IdM models, the full control is with the end user over his/her personal data. Windows Card-Space is a visual metaphor for identity selector for the end-user. Windows Card-Space provides controlling power to the end-users [31]. Windows Card-Space has self-issued cards and has message level security problem while communicating with identity provider. It belongs to the user-centric category, where users may create, delete, or modify identity profiles, known as information cards (i-cards), thus controlling the kind and amount of information revealed to the network. User-centric schemes are capable of providing IdM solutions across different contexts (administrative), as directly end user is responsible for provisioning identity information. A universal mechanism or global identifiers are required to support interoperability in case of user-centric IdM. OpenID [16] is another user-centric platform which is consistent with four layers- a) Identifiers, b) Discovery, c) Authentication, and d) Data Transport. The process, to be completed, involves three different entities- the end –users, the relying party, and the identity provider, same as the identity meta-system. OpenID allows relying party to redirect the client to the identity provider for authentication at the identity provider site thus violating user control. The second problem with OpenID is that the URL that is used to identify the subject is recyclable. Since OpenID permits URL based identification, it brings the issue of privacy.

3.3.2 Device-centric Identity Management

Today the concept of identities for devices/things is in its infancy and when things have identities; it is mostly used for identifying things for inventory, and authentication purposes (e.g. RFID Tags, MAC-IDs, etc). In the future, users will be interacting with things that surround them in a multitude of different ways, for which current identities for things are inadequate. Consider for a moment, how a user can attach a device available publicly to his/her personal space of devices for a short time? How can he/she trust this device? How will this thing access his/her personal information? The identity possessed by the device will form the backbone on which answers to this question can be found.

Higgins [32] is a software infrastructure that supports consistent user experience that works with digital identity protocols, e.g. WS-Trust, OpenID. The main objective of the Higgins project is to manage multiple contexts, interoperability, define common interfaces for an identity system. The Higgins framework does not provide support for quantitative measure identity strength and lacks the fulfillment of defining strength of identity. Higgins also has the inherent weakness of the security vulnerability. The OAuth 2.0 [33] defines a rich set of communication protocols for the authorization of resource accesses. The OAuth authorization framework enables a third-party client to access a resource on behalf of an entity. It is simple framework and the application which runs on the devices with limited capabilities can also take benefit of this for authorization.

3.3.3 Hybrid Identity Management

Hybrid IdM deals with hybrid identities like user as well as device identities. In the emerging area of cloud computing, IdM in the hybrid cloud need to deal with identities of both user and devices/services. In federated IoT environment, hybrid identities and its management is more critical wherein power is delegated to providers and organizations for IdM. Liberty Alliance [15] is federated solution for guaranteeing interoperability, supporting privacy, promoting adoption for its specifications, and provides guidelines. The Liberty Alliance Project lacks from defining strength of identity. It is a framework according to which domains that belong to a federation may exchange identity information about their users and devices using federated identities. Athens [34] is another identity and access management service which provides high

72 *Identity Management Models*

level functionalities for user profile management and SSO services to protected resources.

Shibboleth is a federation infrastructure based on SAML and web redirection with a single sign on mechanism in order to share resources. The Identity Provider is composed by the single sign on service, Inter site transfer service, authentication authority, and attributes authority and artifact resolution service [17]. Shibboleth does not comply with directional identity. There are cases where it is necessary to implement uniquely identifiable identities and also directional. Shibboleth fails to offer such support. One more crucial area where Shibboleth fails to protect itself is against the susceptibility of the security of the whole system be broken down by an evil third party.

Table 3.2 summarizes the areas of IdM covered by various technologies. Comparison is based on following six IdM technologies.

- OpenId [16] – Distributed identities
- Liberty Allince [15] – Trust relations
- Card-Space [31] – Authentication framework
- Shibboleth [17] – Attribute exchange
- Higgins [32] – User and machine Ids
- OAuth 2.0 [33] – Authorization framework

However, managing increasing number of resource constrained devices, mobility of the devices, dynamic network topology and ad-hoc nature makes IoT more vulnerable to security threats and attacks. From Table 3.1, it can be concluded that there is no single technology available that suites a futuristic IoT scenario for many to many (m: n) authentication, multi device SSO, replay attack, lightweight version and device identity with privacy.

Especially, IdM for the IoT should address the following three groups of topics:

Identities: How does the concept of identity and identifiers translate into the world of IoT? How users' interactions with things affect the scope of identities inside IoT? Topics here include identifiers and attributes of users, services and things, new concepts such identity aggregation (e.g. multiple things jointly appearing under a single identity), identity imprinting (e.g. imprinting a temporary identity on a thing), private vs. public things, privacy aspects of identities (traceability, linkability, RFID, NFC, etc.), circles of trust (e.g. things belonging to different owners)

Authentication of an identity: Topics here include methods for authenticating users, services and things, multi-thing single sign-on, authentication

Table 3.2 Comparative summary of the state of the art for identity management

Evaluation Parameters	OpenId	Liberty Alliance	Card-Space	Shibboleth	Higgins	OAuth 2.0
Authentication (1:1)	NO	YES	YES	YES	YES	YES
Authentication (m: n)	NO	NO	NO	NO	NO	NO
Authorization	NO	YES	YES	YES	YES	YES
Technologies	HTTP	SAML	XML	SAML	HTTP	HTTP
Device Identity	NO	PARTIAL	NO	NO	YES	YES
Interoperability	NO	YES	NO	NO	NO	YES
Decentralized	YES	NO	NO	YES	YES	YES
Lightweight	NO	NO	NO	NO	NO	NO
Scalability	YES	NO	NO	YES	YES	YES
Attack Resistant	NO	NO	YES	NO	YES	NO

in case of identity imprinting, authentication in case of uncertainty (trust negotiation, reputation, evidence, etc)

Authorization and attribute exchange: attribute exchange protocols for users and things, selective disclosure of attributes (privacy protection), negotiation, etc.

3.4 Conclusions

We believe that IdM itself is a very big administrative domain, and requires a lot of attention in the future to provide more scalable and complete solutions. It seems that all security protocols are limited by their requirement regarding computational efficiency and scalability due to unbound number of devices. It would be valuable to have more formal analysis for these limitations. The scalability issue in IoT makes IdM of ubiquitous devices more challenging. Forming ad-hoc network, interaction between these nomadic devices to provide seamless service extend the need of new identities to the devices, addressing, and IdM in IoT.

An identity can be considered as being a sum of identifiers and attributes. Multiple Things with own identities can join to create a new single identity; e.g. a public screen and a user private mobile phone may momentarily form a "virtual" personal device, with the user identity and preferences "imprinted". Things can be owned by multiple owners. Things may belong to (multiple) circles of trust; within which they are treated as trusted by other parties (things, services). One major issue is the protection of identities with the IoT environment even though identities are traceable and linkable to the real owners (e.g.: RFID, NFC).

IdM in the context of IoT is discussed in first part of this chapter. Identifiers for objects and association of identifiers with objects are important steps towards identification. Process of identification and proposed identifier format for envisaged IoT is also presented and discussed. IdM is a combination of processes, and technologies to manage, and secure access to the information, and resources while also protecting things' profiles. IdM is an accomplishment of three phase's concern with thing identity in IoT. Identity portrayal is done through the following phases like substance, content and use. There are various identification schemes available in the literature and these schemes have been explored. Analysis of proposed identity schemes concludes that it is dubious to have common identification schemes globally.

Different IdM models in the view of varied identity are discussed in detail to understand how identity flows in the IoT. State of the art in IdM schemes have been also discussed and its evaluation shows that there is no single technology available that suites a futuristic IoT scenario for many to many (m: n) authentication, multi device SSO, replay attack, lightweight version and device identity with privacy.

We conclude that, objectives of IdM in IoT are as follows:
Objectives:

- Conceptual analysis of identities in IoT (identifiers and other attributes of users, services and things, new concepts such as multiple things under a single identity, identity imprinting, private vs. public things)
- Foundations for the system of identifiers for IoT (type and scope of identifiers, grouping of identifiers, etc.)
- Privacy aspects of IoT identities analyzed, solutions proposed for mitigating privacy threats (traceability, linkability, spam, loss of control over personal data, identity theft)
- End-user prospective included in the analysis and foundations

References

[1] Parikshit N. Mahalle, Neeli R. Prasad and Ramjee Prasad, "Novel Context-aware Clustering with Hierarchical Addressing (CCHA) for the Internet of Things (IoT)," In the Proceedings of IEEE Fourth International Conference on Recent Trends in Information, Telecommunication and Computing – ITC 2013, August 01–02, 2013, Chandigarh, India.

[2] EU FP7 Project CASAGRAS, CASAGRAS Final Report: RFID and the Inclusive Model for the Internet of Things, 2009, pp. 43–54.

[3] Lyon, D., "Identifying citizens: ID cards as surveillance," Polity Press, Cambridge, 2009.

[4] R. Moskowitz, and P. Nikander, "Host Identity Protocol (HIP) Architecture," 2006, http://www.ietf.org/rfc/rfc4423.txt.

[5] Parikshit N. Mahalle, Sachin Babar, Neeli R Prasad, and Ramjee Prasad, "Identity Management Framework towards Internet of Things (IoT): Roadmap and Key Challenges," In proceedings of 3rd International Conference CNSA 2010, Book titled Recent Trends in Network Security and Applications - Communications in Computer and Information Science 2010, Springer Berlin Heidelberg, pp: 430–439, Volume: 89, Chennai-India, July 23–25 2010.

[6] Sachin Babar, Parikshit N. Mahalle, Antonietta Stango, Neeli R Prasad, and Ramjee Prasad, "Proposed Security Model, and Threat Taxonomy

for the Internet of Things (IoT)," In proceedings of the 3rd International Conference CNSA 2010, Book titled: Recent Trends in Network Security and Applications - Communications in Computer and Information Science 2010 Springer Berlin Heidelberg, pp: 420–429 Volume: 89, Chennai – India, July 23–25 2010.
[7] Amardeo Sarma, and Joao Girao, "Identities in the Future Internet of Things," In Springer Wireless Personal Communications, Volume: 49, Issue: 3: pp: 353–363, May 2009.
[8] López Tomás Sánchez, Brintrup Alexandra, Isenberg, Marc-André, and Mansfeld Jeanette, "Resource Management in the Internet of Things: Clustering, Synchronisation, and Software Agents," Book Title: Architecting the Internet of Things: Springer Berlin Heidelberg, 159–193. 2011.
[9] W. Heinzelman, A. Chandrakasan, and H. Balakrishnan, "An Application-Specific Protocol Architecture for Wireless Microsensor Networks," In IEEE Transactions on Wireless Communications, Volume: 1, No. 4, pp: 660–670, October 2002.
[10] Tingrong Lu, Yushu Ma, and Yongtian Yang, "Hierarchical Addressing in IP Networks," In Proceedings of International Conference on Communications, Circuits, and Systems, Volume:2, no., pp:1267–1271. Hpng Kong – China, May 27–30 2005.
[11] Chamlee M.E., Zegura E.W. and Mankin A., "Design, and Evaluation of a Protocol for Automated Hierarchical Address Assignment," In Proceedings. Ninth International Conference on Computer Communications, and Networks, Volume., no., pp: 328–333, Las Vegas – NV, October 16–18 2000.
[12] Yinfang Zhuang and Calvert K.L., "Measuring the Effectiveness of Hierarchical Address Assignment," In IEEE Telecommunications Conference, (GLOBECOM 2010), Volume., no., pp:1–6. Florida-USA, December 6–10 2010.
[13] Chandramouli R., and Rose S., "Challenges in Securing the Domain Name System," In Security & Privacy IEEE Journal, Volume: 4, Issue:1, pp: 84–87, January-February 2006.
[14] Qiang Shen, Yu Liu, Zhijun Zhao, Song Ci, and Hui Tang, "Distributed Hash Table Based ID Management Optimization for Internet of Things," In Proceedings of the 6th International Wireless Communications and Mobile Computing Conference, (ACM -IWCMC '10), pp: 86–690, Caen-France, June 28-July 2 2010.
[15] The Liberty Alliance Project - www.projectliberty.org.

References

[16] OpenID – www.openid.net.
[17] The Shibboleth project – www. shibboleth.net.
[18] Web Services Security Specifications Index Page on MSDN. http://msdn.microsoft.com /en-us/library/ms951273.aspx.
[19] 3GPP TS 33.222- Generic Authentication Architecture (GAA); Access to Network Application Functions using Hypertext Transfer Protocol over Transport Layer Security (HTTPS) - http://www.3gpp.org/ftp/Specs/archive/33_series/33.222/.
[20] Federated Identity Management based on Liberty. EU CELTIC project. http://www.celtic-initiative.org/Projects/Celtic-projects/Call2/FIDELITY/fidelity-default.asp.
[21] Service Platform for Innovative Communication Environment. EU FP6 project.www.ist-spice.org/.
[22] The SWIFT (Secure Widespread Identities for Federated Telecommunications) Project, 2008: www.ist-swift.org/.
[23] Bayu Anggorojati, Parikshit N. Mahalle, Neeli R. Prasad, and Ramjee Prasad, "Secure Access Control and Authority Delegation based on Capability and Context Awareness for Federated IoT," In Internet of Things and M2M Communications Book, River Publications, May 2013, Edited by: Fabrice Theoleyre (University of Strasbourg, theoleyre@unistra.fr) & Ai-Chun Pang (National Taiwan University, acpang@csie.ntu.edu.tw).
[24] Hirsch, P.M., "Exercise the power of the World Wide Web," Computer Applications in Power, IEE, vol.8, no.3, pp.25, 29, Jul 1995.
[25] Long Jin; Yang Chen; Tianyi Wang; Pan Hui; Vasilakos, A.V., "Understanding user behavior in online social networks: a survey," Communications Magazine, IEEE , vol. 51, no. 9, pp.144, 150, September 2013.
[26] M. Nam Ko et al., "Social-Networks Connect Services," Computer, vol. 43, no. 8, Aug. 2010, pp. 37–43.
[27] Bemers-Lee, T, Uniform Resource Identifiers (URI) Syntax, IETF RFC 2396, 1998.
[28] Lampropoulos, K.; Denazis, S., "Identity management directions in future internet," Communications Magazine, IEEE, vol. 49, no. 12, pp. 74, 83, December 2011.
[29] Torres, J.; Nogueira, M.; Pujolle, G., "A Survey on Identity Management for the Future Network," Communications Surveys & Tutorials, IEEE, vol. 15, no. 2, pp. 787, 802, Second Quarter 2013.

[30] Jensen, J.; Jaatun, M. G., "Federated Identity Management - We Built It; Why Won't They Come?," Security & Privacy, IEEE, vol. 11, no. 2, pp. 34, 41, March-April 2013.
[31] Microsoft Corporation, "Windows CardSpace," 2006, (accessed 2008-08-15). http://cardspace.netfx3.com/.
[32] Higgins: http://www.eclipse.org/higgins/.
[33] D. Hardt, The OAuth 2.0 Authorization Framework, IETF, RFC 6749.
[34] Athens - http://www.athensams.net

4

Identity Management and Trust

4.1 Introduction

In the vision of ubiquitous computing, the activities of daily life are supported by a multitude of heterogeneous, loosely coupled computing devices. The support of seamless collaboration between users as well as between their devices can be seen as one of the key challenges for this vision to come true. Adequate management of identities in IoT is crucial to provide security, and to improve efficiency. IdM requires an integrated and often complex infrastructure where all involved entities must be trusted for specific purposes depending on their role. The variety and complexity of the trust relationships required in the various IdM models can cause confusion for stakeholders. Satisfying the expected trust requirements is also associated with a cost. By integrating the physical world with the information world, and providing ambient services, and applications, ubiquitous networks allow users, devices, and applications in different physical locations to communicate seamlessly with one another. However, the decentralized and distributed nature of IoT face challenges on trust management, access control and IdM [1]. The classical and centralized mechanism does not suffice because of the distributed nature of the device-to-device communication. Without the effective IdM, and access control, the benefits of ubiquitous networks will be limited. For example, in ubiquitous healthcare if access control and IdM is not guaranteed, it can lead to leakage of medical data.

Identity and trust have been indicated as highly important aspects in the context of IoT networks [2]. Evaluation of the state of the art in identity shows that there is no common and universally accepted solution for IdM in the IoT. Due to continuous growth and wide range of applications with the convergence of RFID, sensor networks and distributed applications, there are more challenges for IdM. Existing IdM solutions and different identification schemes addresses a specific aspect of IdM but does not provide overall and complete approach for identity problem. Multiple approaches to IdM problem have resulted to the problem of identity anarchy [3]. Identity anarchy problem refers to the identity theft, identity corruption making identity problem more complex. One of the most critical issues in the IoT is trust which can be summarized in two points, first is whether IoT devices have trust to each other, and second is how access is granted depending on the trust? According to [4], the research areas of IdM and privacy should include device authentication, new concept of identities, trust-based access control and access control.

Trust plays an important role in large distributed systems such as IoT networks. Increasing number of applications, services, sensors, devices raises the issue "whom to trust and whom not to". Everyday things are globally connected and managing increasing number of things requires scalable and efficient trust management solution. Due to increasing number of connected things in IoT, energy, ubiquitous network access, secure user interaction increases the complexity of operation. Designing a trust management model to provide trust in IoT is thus an important step towards achieving the privacy and security of entities in such a decentralized, distributed and mobile space. In this context, without human judgment, the challenge for devices is to be able to distinguish other peers' identities and behaviours autonomously. Trust-based access control is a promising approach where identity and trust are encapsulated together. Trusted identity enables to grant customized access to IoT networks based on device identity. IdM requires the establishment of trust between communicating devices and as described by the International Telecommunications Union, an entity can be said to "trust" a second entity when it (the first entity) has reason to assume that the second entity will behave exactly as the first entity expects. A relation between identity and trust is also presented in [5–7].

The trust provides devices with a natural way of judging another device, similar to how we have been handling the security, and access control in human society. Once a trust relationship is established between the two devices after communicating, and collaborating for a certain time, it will help in influencing the future behaviors of their interactions. When devices trust each other, they prefer to share services, and resources to a certain extent. Trust management allows the computation, and analysis of trust among devices to make suitable decisions in order to establish efficient, and reliable communication among devices [8].

4.1 Introduction 81

4.1.1 Motivation

To achieve access control for IdM, relation between trust and access control plays an important role. In IoT, trusted devices are only the authorized object to access resources. The access credentials can be exchanged, and evaluated mechanically using trust negotiation. Binding trust and identity together addresses important issues like privacy protection and identity theft. Using efficient trust model, scalability can be achieved which is the one of the most important design issues in the context of IoT. Adequate management of identities in IoT is crucial to provide security and access control. IdM requires an integrated and often complex infrastructure where all involved devices must be trusted for specific purposes. The trust plays a crucial role, and is recognized as a major risk factor in IdM. Designing a trust management model to provide trust in IoT is thus an important step towards achieving the security and access control of devices in such a decentralized, distributed and mobile space. In this context, without human judgment, the challenge for devices in IoT is able to distinguish other peers' identities, behaviors, and access control autonomously. As an example, a user might want to send a sensitive document from his/her PDA to a public printer directly via a transient, peer to peer Bluetooth radio link without gaining access to a centrally administered intranet. In such ad-hoc interactions, the participating devices do not always have membership within a network. Each device will have to assume that arbitrary device can establish direct, ad-hoc communication with it. The device may simultaneously provide services to more than one network. Consequently, every device becomes a potential gateway to leak information across the network perimeters. This makes it difficult to establish, and defend the borders of IoT.

Rather than depending upon the network topology to establish trust, the device itself must be involved to enforce trust-based access control. Building upon our earlier example, if a user wants to wirelessly print a document from a PDA on one of the five available public printers at an airport lounge, it is difficult to establish with certainty that the device is talking only to that specific physical printer with proper access control in place, and not some other device in the vicinity. Consider another scenario where Mark is technophile, and by profession a salesman. His job requires business travels across the globe. He can access information and services both private, and professional through his latest devices developed for IoT. On one of Mark's business trips, he enters the airport, gets an alert on his smart device showing the different services available at the airport e.g., a guided map of the airport, the current waiting time in the security check area, airline services etc. He chooses to check the current waiting time, and is informed by a device in the airport that on an average it takes half an hour to clear security. At the check-in desk, another alert informs him that due to a technical snag his flight is delayed by a couple of hours, and lunch e-vouchers are provided by the airlines where alerts based on personal information are made available to the airline thing. Mark checks his email. The company device can access services subscribed by his company worldwide. A fast internet connection to access services and the office is available with the company's subscription where there is provision of services on being part of a group. The sscenario presented above shows that there is a need of scalable trust management model for access control in IoT.

4.1.2 Trust Management Life Cycle

The trust is a particular level of the subjective likelihood belief with which an entity will perform a particular action, before one can monitor such action, and in a context in which it affects our own action. The trust is context-dependent, dynamic, and non-monotonic parameter. The trust management was first coined by Blaze [9] in 1996 as a coherent framework for the study of security policies, security credentials, and trust relationships. The mechanism that deals with the evaluation, collection, and propagation of trust is referred to as trust management. There are three types of trust viz:

a) **Interpersonal** trust represents entity-based, and context specific trust.
b) **Structural** trust represents a system within which the trust exists.
c) **Dispositional** trust represents a trust which is independent of entity, and context.

There are different trust management approaches and generic trust management life cycle is shown in the Figure 4.1. In a nutshell, any trust management model comprises of four phases of trust calculations as:

- Negotiation – Trust establishment between new devices
- Collection – Collecting trust scores of individual device in IoT
- Evaluation – Deals with the trust evaluation based on some fuzzy, or non-fuzzy rules, and some evaluation policies
- Propagation – Transfer of trust score to other devices, and in turn delegating other details like access rights etc.

There are different ways to infer trust like probabilistic method, Bayesian method and inference method. Probabilistic and Bayesian methods [10] are based on the principle of uncertainty and needs strong methods of proof to

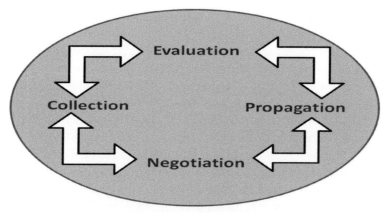

Figure 4.1 Trust management life cycle

claim that proposed or presented solution would work with trust management along with authentication and access control. Thorough survey has been done on trust management models for wireless communication [11, 12]. Survey shows that there could be individual level trust model or system level trust model. In [13], trust is defined as confidence in exchange partner's reliability and integrity. This confidence provides basis to believe in reliability and integrity. [14, 15] examined how privacy is related to trust and presented privacy paradox but especially address web scenarios. Efficient trust management establishes stronger form of access control for ubiquitous objects. Trust management results into functional system in which an access request is accompanied by set of credentials which together constitute a proof as to why the access should be allowed.

4.1.3 State of the Art

In the Internet and web computing, there are various approaches which do not only address the trust issue but also privacy. Existing standards like SMIME [16], OpenPGP [17], and Internet X.509 Public Key Infrastructure [18] allow transmission of information over an untrusted channel by ensuring integrity and also ensures the recipient that the information is originated from intended origin. PGP is the first initiative towards Web of trust where communicating entities can express degree of trust in each other. However, most of the developed approaches are based on increasing user trust perceptions, than involving users to make correct trust decisions. It is also possible to create fake identities and these solutions become the problem [19]. It must also be noted that, these internet-based solutions are user-driven and trust between devices is missing.

The access control mechanism based on the trust calculations using fuzzy approach is presented in [20] where access feedback is used for access control. This scheme is not suitable for distributed nature of IoT. A calculus for granting access is introduced in [21] where notion of control is introduced to state whether principal Pi is trusted on a concept, or Pi is not trusted. In [21], ranking is also introduced in order to express the predicate that principal P_i is stronger that P_j as $P_i \Rightarrow P_j$. The formal model of trust is presented in [22], and focused on the aspects of trust formation, evolution, and propagation based on domain theory. A fuzzy logical system to deal with trust management is presented in [23] where a complete system for trust management is discussed. Another interesting approach of trust management is proposed in [24] in which a novel framework is presented with the Principal (Entities that can make or authorize request), and Authorization as main elements. Probabilistic trust management approach for pervasive devices in terms of device-to-device interactions, and security analysis is presented in [10]. Comparison of this probabilistic approach is made with deterministic approach, and proved that probabilistic approach is better in terms of performance, and security of interaction with dynamic adaptation to changes in environment. Mutual trust establishments based on expected utility with experimentation results are presented in [25]. The socio-cognitive trust model using fuzzy cognitive maps is presented in [26]. Trustworthiness based on beliefs, and its computational model is presented in [26]. Mathematical framework of trust for cognitive radio networks is presented in [27] as cognitive network is one of the multi-hop heterogeneous wireless networks. A novel approach of integrating trust management with access control is also presented in [28] where structured query language-based syntax with algorithm for end-to-end security is presented. Trust requirements in IdM are presented in [29]. Scalable trust management protocol with the emphasis on social relationships is presented in [30]. Aggregated trust based on direct and indirect observation is presented, and performance comparison for service composition is also presented in [30].

Majority of the literature presents individual level trust model and there is a need of hybrid trust model with trust score calculation. Individual level and hybrid trust model cannot address security issues as required. Also communication overhead reduction and generic framework has not been addressed. There is also a need of explicit trust model which will address trusted authentication and access control for ad-hoc wireless communication. It must however be noted that all of the above models serve their purpose in their own domains, which are probably sufficient for the current world of computing, and it must again be stressed that the fuzzy approach for trust management can be indeed a new requirement.

4.2 Identity and Trust

The risk of privacy violation, omnipresent network, and reliability issues challenges trust in the IoT. Therefore, IoT networks and applications should be trustworthy. Trustworthy interactions should take care of data and identity protection against attacks and theft, privacy and security management. As pointed out in [31], "the failure to enhance trust may result in suspicion and eventual rejection of new technology". In order to be trusted, the IoT must provide methods for secure and reliable communication between devices, which can be clearly and mutually authenticated. In order to provide expected level of assurance, identity of the things must be protected. The trusted IoT must take into account security, identity assertion and management. Trust and identity research shows that, trust is a primary design element at every layer of IoT architecture. Existing research in trust and identity presents number of challenges like trustworthiness of IdM schemes, robust trust/IdM, privacy-aware trust/IdM and essentially resource constrained nature of the IoT. Trust in identity authentication is established in the IoT by following assertion: The entity performing authentication is presented with information that only the entity being authenticated is able to provide. This information is referred to as proof of possession (POP) of identity. The authenticating entity establishes trust in this process through a secure verification of the presented proof [32]. Due to distributed nature of the IoT, the scope of identity goes beyond the boundaries of enterprise network and trust play a key role. There are many ways of binding identity and trust together in the computing and they vary based on the method of trust calculation. It is evident that the trust in identity is an entry to all trust management models in the context of IoT [33, 34]. There are different trust paradigms and main paradigms for identity trust are third party approach, public key infrastructure and attribute certificates and are presented in next section.

4.2.1 Third Party Approach

In the local identity model, object identity is used in the repository/registry of every network. In such models, shared secret used for building trust is different for different systems to avoid compromise. This increases the complexity of the system due to management of multiple secrets. Identity is local for the local scope in which it is identified and this results into complex system and lack of scalability. Due this limitation of local IdM model, third party approach of identity and trust management has emerged. In third party approach, single entity (E.g. host/cloud/network) is designated as trusted by all the stakeholders in the network, such as users, computing agent, and applications. Entity details at third party include identity information of all the stakeholders. There are various applications of third party approach like Kerberos and key distribution center (KDC). Trust is build based on the shared secrets between every entity and the third party identity establishment service. There are two identity establishment paradigms based on the third party approach:

- **Implicit identity establishment:** Mutual identity establishments of entities to each other via trusted third party. Cryptographic approach is used with the help of shared secret and third party.
- **Explicit identity establishment:** Third party authentication service is invoked for authentication explicitly.

Computing analysis shows that, direct identity establishment between users and services requires managing n * m secrets wherein, in implicit third party identity establishment scheme requires to manage m*n secrets.

The main problems associated with third party approach are [35]:

a. *Loss of control:* Trust management scheme is located at third party and underlined network has to rely on third party for security and privacy.
b. *Lack of trust*: Trusting on third party requires assumption that the third party will perform or act as it is expected. Additional monitoring or auditing capabilities would require being in place to increase the trust level.
c. *Multi-tenancy:* As a single third party is associated with multiple networks for providing trust, these networks can have multiple goals which could be conflicting. Some degree of separation between these networks would be useful.

4.2.2 Public Key Infrastructure

RFC 2822 defined Public Key Infrastructure (PKI) as the set of hardware, software, users, and digital certificate management based on public key cryptography. Objective of PKI is to enable secure and efficient management of public keys. One of the important features of public key cryptography is that the trust can be established without having to share secrets and public keys are intended to be universally accessible.

In the Internet computing, public key establishment is defined by X.509 digital certificates issued by trusted third party i.e. Certification Authority (CA) [36]. X.509 certificate represents crypto binding between public key and identity. Assurance of public key in the certification process is provided by PKI. Communicating parties relying on public keys have their trust on third party i.e. CA. Before public key is distributed, trusted CA uses its own private key to digitally sign the entities' key, which is then sent to the pool of public key. Trust model in PKI is based on degree of assurance in public key certificate of issuing trusted CA. The public key of the issuing CA derived from its public key certificate is used to verify the digital signature of that CA in the entities' public key certificate. Assurance in public key certificate of the issuing CA, verification of signature establishes trust in binding public key with the identity of the entity that holds the public key certificate. Figure 4.2 depicts the summary of different PKI trust topologies.

4.2.3 Attribute Certificates

Attribute is defined as a particular type of information to identify entity or object. The component of attribute to indicate class of information is referred as attribute type and attribute value is particular instance of information indicated by attribute type. In the context of security, attribute is defined information associated with a key that is not used in cryptographic algorithm. Attribute Certificate (AC) is a digital certificate which encapsulates a set of data items (other than public key) directly to entity name or to public key certificate of another entity. AC is a data structure digitally signed by Attribute Authority (AA) which binds some attribute values with the identification information of its holder. AA is a trusted CA that issues AC and CA may also be AA. PKC is signed and issued by CA which binds identity of the holder with public-private key pair. The difference between AC and Public Key Certificate (PKC) is presented in Table 4.1.

AC mainly contains set of entity attributes, a validity period, and signature certifying the integrity of the AC and identity establishment of its issuing authority. All the attributes of AC are encapsulated in the *AttributeCertificate-Info* data type. Signature information of the AC is not included in this. Format of AC in Abstract Syntax Notation (ASN).1 is shown Figure 4.3 below. As shown in Figure 4.3, AC contains mainly:

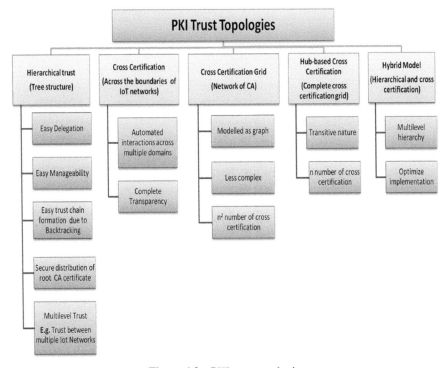

Figure 4.2 PKI trust topologies

1. Binding information: Helps to enable AC verifier to confirm trust.
2. Attribute Information: Contains sequence of uniquely identifiable attributes.

4.2.3.1 Binding information

AC binding information establishes association between AC, its issuer and its holder. It contains following fields for binding.

Issuer: The issuer of AC is represented by equivalent X.500 specific name. All AC issuer have nonempty distinguished names. AC verifier maps the issuer name to a PKC for confirming trust.

Holder: When AC is being used in an authenticated message of a handshaking session in which identity establishment is based on the X.509 PKC (E.g. TLS/SL), holder field contains the holder's PKC serial number and issuer. This binding helps to establish authenticated security context in which the AC can be used to perform authorization checks.

Serial Number: This is unique serial number assigned to AC.

4.2 Identity and Trust   89

Table 4.1 Difference between AC and PKC

AC	PKC
Issued by AA	Issued by CA
AC contains no public keys.	X.509 contains public keys.
AC contains set of attributes of its holder.	X.509 does contain set of attributes of its holders.
Validity period of AC is less.	Validity period of PKC is likely to be more than the lifetime of AC.
AC is dynamic in nature and is constantly subject to change.	PKC is likely to unchanged and valid for long period of time.
E.g. Capability token	X.509 certificate

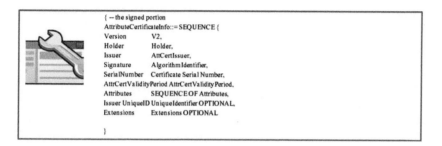

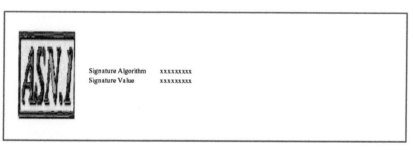

Figure 4.3   X.509 v2 AC

4.2.3.2 Attribute information

This information comprises set of uniquely identifiable attributes with a collection of key value pairs. Privilege attributes are important for access control and it form the base of AC. The number of AC attributes is standardized in order to maintain interoperability across various security domains. Some of them are briefed below:

Service authentication information: It is used to identify the AC holder to underline service by name. The holder's identity and password to legacy applications is communicated through this attribute. Encryption scheme can also be used for the protection of authentication information.

Charging Identity: Identity used by the AC holder for charging purpose is represented by this attribute. Example of this attribute is billing service.

Role: It presents the role of AC holder and additionally it specifies the name of authority issuer.

Clearance: It carries clearance information associated with the AC holder for multilevel security.

4.3 Web of Trust Models

Authenticity of binding between public key and its owners is established by Web of Trust (WoT). WoT presents decentralized model more suited in the context of IoT than centralized trust model of PKI based on CA. It is distributed trust model of decentralized PKI. SSL-based web model between two HTTP endpoints (client and web server) uses simple trust model where user can select any root CA which seems more trustworthy. Entities in this trust model maintain multiple root CA certificates in its local space and validating these certificates requires finding trust path to one of the trusted CA. Hierarchical trust models are easily represented as a tree of nodes, but the WoT is closely related to the human notion in a way how people determine the trust in their daily life. As people traverse through life and come across new people, they look towards the already trusted people to decide whether they should trust new people. There are various problems in the trusted exchange of public keys in public key cryptography. WoT is a good approach where entity holds the responsibilities for identifying and authenticating each other and then swap their respective keys. PKI is one alternative approach where entities are identified and authenticated by trusted CA. Instead of communicating entities themselves involved in exchanging keys, entities obtain keys from one or more CAs in the form of digital certificates.

Pretty Good Policy (PGP) is the trust model based on WoT initially developed by Philip Zimmermann as an email encryption program [37]. PGP uses public key encryption for the distribution of secret encryption keys. Trust scheme used in PGP is known as PGP WoT which is based on the discretionary trust of individuals without the concept of authoritative entity that certifies public keys in PGP. An entity generates a public-private key pair that binds to unique identifier like email address and distribute to other entities or key distribution services. Each entity maintains a set of public keys of other entities which are trustworthy. Trust in PGP model is not transitive which means that A trusts B as an introducer and in turn B trusts C does not necessarily establish that A trusts C.

A signs B implies "A has some level of trust in the authenticity of B". If there is a chain of trust from a key that I trust (my key) to a key that I don't know anything about, some sense of trust can be established. The WoT is a directed graph of trust relationships. The PGP web of trust can be modeled by a directed graph $G = (N, E)$ where the set of nodes N represents the collection of entities participating in a PGP web of trust, and edge $e \, \varepsilon \, E$ from entity A to entity B represents the fact that A trusts the public key of B. Figure 4.4 presents scenario of WoT and it can be seen that paths through the WoT is asymmetric and path diversity is good.

Emerging mechanisms for the exchange of security constructs in the context of web or IoT computing are:

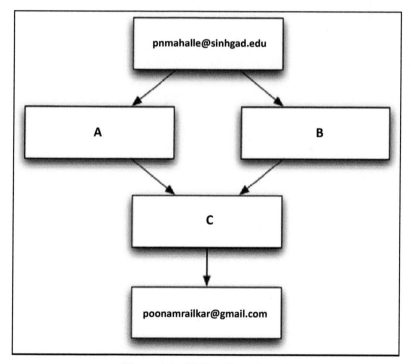

Figure 4.4 WoT as directed graph

1. Web services security
2. Secure Assertion Markup Language
3. Trust

4.3.1 Web Services Security

Service is a well defined function that does not depend on the state of the other services. Consumer needs to know how to call the service and what to expect in response. Two softwares can communicate with each other using SOA (services oriented architecture). Web services (WS) is an implementation of SOA. Service provider publishes its service description on a directory. Service consumer performs queries to the directory to locate a service and finds out, how to communicate with provider. Service description is written in a special language called web services description language (WSDL). Messages are sent and received in a special language called simple object access protocol (SOAP).

Web service architecture is depicted in the Figure 4.5.

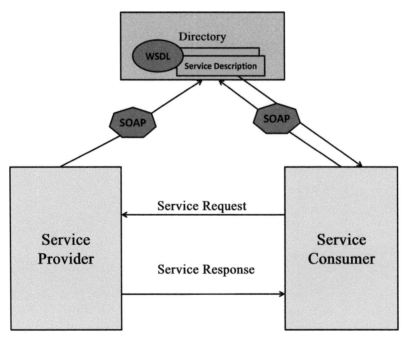

Figure 4.5 Web service architecture

4.3.1.1 WS-Security
Web Services Security (WS Security) is a specification that defines how security measures are implemented in web services to protect them from external attacks. It is a set of protocols that ensure security for SOAP-based messages by implementing the principles of confidentiality, integrity and authentication. Because Web services are independent of any hardware and software implementations, WS-Security protocols need to be flexible enough to accommodate new security mechanisms and provide alternative mechanisms if an approach is not suitable.

Since SOAP- based messages traverse multiple intermediaries, security protocols need to be able to identify fake nodes and prevent data interpretation at any nodes. WS-Security combines the best approaches to tackle different security problems by allowing the developer to customize a particular security solution for a part of the problem. For example, the developer can select digital signatures for non-repudiation and Kerberos for authentication. All data related to security is added as part of the SOAP header. Therefore, a considerable overhead is imposed on the SOAP message formation when security mechanisms are activated.

4.3.1.2 WS-Security SOAP header
The developer is free to choose any underlying security mechanism or set of protocols to achieve his/her goal. Security is implemented using a header which consists of a set of key-value pairs where the value changes appropriately with changes in the underlying security mechanism used. This mechanism helps to identify the caller's identity. If a digital signature is used, the header contains information about how the content has been signed and the location of the key used to sign the message.

Information related to encryption is also stored in the SOAP header. The ID attribute is stored as part of the SOAP header, which simplifies processing. The timestamp is used as an additional level of protection against attacks on the message integrity. When a message is created, a timestamp is associated with the message indicating when it was created. Additional timestamps are used for the expiry of the message and to indicate when the message was received at the destination node.

4.3.1.3 WS-Security authentication mechanisms
- Username/Password approach
- X.509 approach
- Kerberos

- Digital Signature
- Encryption

4.3.2 SAML Approach

As stated earlier, extension of IdM to multiple security domains is referred as federated IdM. It includes autonomous internal networks, external networks and third party applications and services. The objective of federated IdM is to provide sharing of digital identities so that objects can be authenticated a single time and can access applications and services across multiple security domains in the context of IoT. As the IoT is distributed in nature, no centralized control is possible, federated IdM is a need of IoT. Identity provider, service provider and users are the key entities in federated IdM. For example, IoT device may log onto own corporate intranet and will be authenticated to perform authorized activities and access appropriate services on the underlined corporate intranet. The same IoT device is then able to access health benefits of the user from external healthcare service provider without having to re-authenticate.

Main functions of federated IdM are:

- Providing agreement, standards and technologies for identity portability
- Exchange of identity attributes across multiple security domains
- Entitlements of multiple devices/users across multiple domains and applications
- Providing SSO
- Identity mapping across multiple security domains

Federated IdM uses number of standards for the secure identity exchange across multiple security domains. In the sequel, one IoT network or organization issues some security ticket for the access which can be processed by communicating networks. In this view, another task to be carried out in identity federation is to define these tickets in terms of contents and format and providing some rules for exchanging security tickets and performing the functions mentioned above. Key standard in federated IdM is SAML which defines the exchange of security information between communicating partners. SAML is basically intended for expressing trust and identity constructs. SAML is an XML standard for exchanging authentication and authorization data between entities. SAML is a product of the OASIS Security Services Technical Committee. SAML is built upon the following technology standards:

- Extensible Markup Language (XML)

- XML Schema
- XML Signature
- XML Encryption (SAML 2.0 only)
- Hypertext Transfer Protocol (HTTP)
- SOAP

A SAML specification includes:
- Assertions (XML)
- Protocols (XML + processing rules)
- Bindings (HTTP, SOAP)
- Profiles (= Protocols + Bindings)

Figure 4.6 presents SAML components and are summarized below:
- *Assertions:* Authentication, Attribute and Authorization information
- *Protocol:* Request and Response elements for packaging assertions
- *Bindings:* How SAML Protocols map onto standard messaging or communication protocols
- *Profiles:* How SAML protocols, bindings and assertions combine to support a defined use case

4.3.3 Fuzzy Approach for Trust

The modern concept of uncertainty was presented by Lotfi A. Zadeh [38]. He introduced a theory of fuzzy sets where the boundaries are not perfectly defined where the membership is a matter of degree. The concept of fuzzy sets not only provides the meaningful and powerful representation of measurement uncertainties, but also the meaningful, and powerful representation of vague concepts expressed in a natural language, whereas crisp sets are defined by sharp boundaries. In uncertain environments like IoT, fuzzy approach for trust calculations is more appropriate to quantify, and evaluate device behavior, and in turn access control rules. The trust management system should address the questions like kind of authorization device A has on device B.

We recommend Mamdani-type [39] fuzzy rule-based model, which deals with the linguistic values of KN, EX, and RC where vagueness is associated. The output of this model is represented by a fuzzy set. To validate the performance of the model, fuzzy value of the trust can be converted in to a

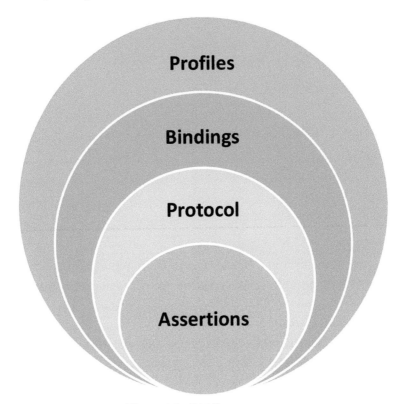

Figure 4.6 SAML components

crisp value by defuzzification methods. The Mamdani scheme is a type of fuzzy relational model where each rule is represented by an If–Then relationship. Mamdani type fuzzy If-Then Rule is written as shown in Equation (4.1):

$$If\ X_1\ is\ A_{1r}\ and\ \ldots\ldots\ and\ X_n\ is\ A_{nr}$$
$$then\ Y\ is\ B_r \qquad (4.1)$$

Where A_{ir} denotes the linguistic labels of the ith input variable associated with the rth rule (i = 1, ..., n), and B_r is the linguistic label of the output variable, associated with the same rule. Each A_{ir} and B_r has its representation in the membership function μ_{ir} and γ_r respectively. The fuzzy output F(y) of the system has the following form as shown in Equation (4.2)

$$F(y) = \bigcup_{r=1}^{R} ((\bigcap_{i=1}^{N} \mu_{ir}(x_i)) \bigcap \gamma_r \qquad (4.2)$$

The complexity in measuring trust score and predicting trustworthiness in service-oriented IoT networks is most promising and leads to many problems. These include how to quantify the capability of individual devices in the trust dynamics and how to assign concrete level of trust in device-to-device communication. Also trust relationship in IoT environment is hard to ascertain due to uncertainties involved. The benefits of fuzzy trust calculations are as follows:

- Inferences using fuzzy approach can easily quantify uncertainties for measuring the level of trust in uncertain IoT environment.
- It is easy to develop membership function and inference rules for different trust relationship using fuzzy approach.
- Another advantage of fuzzy approach as compare to the other approaches is that it can handle incomplete and imprecise inputs in decentralized environment where resource owners usually do not have complete and precise inputs.
- Fuzzy approach is flexible, intuitive knowledge-based tool which is easy for computation and validation.

4.4 Conclusions

Due to the distributed nature of IoT, the scope of identity is spread across the federated networks and the Internet. Identity establishment achieves verifiable security context, success and strength of the security context depends on the IdM. Trust is the main component that establishes and maintains the identity establishment in the context of IoT. Identity and trust is crucial to the success, and full thrives of IoT communication, especially for the device-to-device communication. This chapter presented the study of different trust management models with their advantages, limitations, and introduced how identity and trust works together. A relationship between access control, and trust along with trust management life cycle in the context of IoT is presented in this chapter.

In the sequel, key paradigms in identity trust computing are presented. In the first part, third party approach to identity and trust is discussed. Problems associated with third party approach for trust management are also discussed. In the second part, PKI and its objectives are elucidated. Different PKI trust topologies and their comparison is discussed to understand its application. In the third part, attributes certificates, its representation and its use are explained. The difference between AC and PKC with example shows that even when the elegance, strength, and soundness of one method or another can be apparent, the intent is to explore the elements of trust that characterize each method.

Web of trust models plays important role in Web 3.0. When all the ubiquitous devices are connected to the Internet, authenticity of binding between public key and its owners is established by WoT. WoT presents decentralized model more suited in the context of IoT than centralized trust model of PKI based on CA. Emerging mechanism for the exchange of security constructs like web service security, SAML as well as fuzzy approach for trust computing are then presented and discussed at the end of this chapter.

References

[1] R. Khare, and A. Rifkin, "Trust Management on the World Wide Web," In Elsevier Journal of Computer Networks, and ISDN Systems, Volume: 30, Issues: 1–7, pp: 651–653, April 1998.

[2] Constantijn van Oranje, Joachim Krapels, Maarten Botterman, and Jonathan Cave, "The Future of the Internet Economy: a Discussion Paper on Critical Issues," Technical report, RAND Europe, 2008.

[3] Identity Grid. Technical report, Sun Microsystems, 2004.

[4] Zeta Dooly, Jim Clarke, W. Fitzgerald, W. Donnelly, Michel Riguidel, and Keith Howker, "SecureIST D3.3: ICT Security and Dependability Research beyond 2010: Final strategy," Technical report, SecureIST EU Project, 2007.

[5] Ghazizadeh, E.; Zamani, M.; Ab Manan, J. L.; Khaleghparast, R.; Taherian, A., "A trust based model for federated identity architecture to mitigate identity theft," Internet Technology And Secured Transactions, 2012 International Conferece For, vol., no., pp. 376, 381, 10–12 Dec. 2012.

[6] Lingli Zhao; Shuai Liu; Junsheng Li; Haicheng Xu, "A dynamic Access Control model based on trust," Environmental Science and Information Application Technology (ESIAT), 2010 International Conference on, vol.1, no., pp. 548, 551, 17–18 July 2010.

[7] Qun Ni, Elisa Bertino, and Jorge Lobo. 2010. Risk-based access control systems built on fuzzy inferences. In Proceedings of the 5th ACM Symposium on Information, Computer and Communications Security (ASIACCS '10). ACM, New York, NY, USA, 250–260.

[8] Shunan Ma, Jingsha He, Xunbo Shuai, and Zhao Wang, "Access Control Mechanism Based on Trust Quantification," In IEEE Second

International Conference on Social Computing (SocialCom), Volume: Issue: pp: 1032–1037, Minneapolis-USA, August 20–22 2010.
[9] M. Blaze, J. Feigenbaum, and J. Lacy, "Decentralized Trust Management," In Proceedings of the IEEE Symposium on Research in Security and Privacy, pp: 164, Oakland - CA, May 1996.
[10] M.K. Denko, T. Sun, Probabilistic trust management in pervasive computing, in: Proceedings of the IEEE/IFIP International Conference on Embedded and Ubiquitous Computing (EUC'08), December 17–20, Shangai, China, 2008, pp. 610–615.
[11] Han Yu; Zhiqi Shen; Chunyan Miao; Leung, C.; Niyato, D.; "A Survey of Trust and Reputation Management Systems in Wireless Communications," Proceedings of the IEEE, vol. 98, no. 10.
[12] Esch, J., "Prolog to A Survey of Trust and Reputation Management Systems in Wireless Communications," Proceedings of the IEEE, vol. 98, no. 10, pp. 1752–1754, Oct. 2010.
[13] R. M. Morgan and S. D. Hunt, "The commitment-trust theory of relationship marketing," J. Marketing, vol. 58, no. 3, pp. 20–38, July 1994.
[14] G. R. Milne and M.-E. Boza, "Trust and concern in consumers' perceptions of marketing information management practices," J. Interact. Marketing, vol. 13, no. 1, pp. 5–24, 1999.
[15] P. A. Norberg, D. R. Horne, and D. A. Horne, "The privacy paradox: Personal information disclosure intentions versus behaviors," J. Consumer Affairs, vol. 41, no. 1, pp. 100–127, 2007.
[16] B. Ramsdell. S/MIME Version 3 Message Specification. Technical report, RFC 2633, June 1999.
[17] J. Callas, L. Donnerhacke, H. Finney, and R. Thayer. OpenPGP Message Format. Technical report, RFC 2440, November 1998.
[18] R. Housley, W. Polk, W. Ford, and D. Solo. Internet X.509 Public Key Infrastructure Certificate and Certificate Revocation List (CRL) Profile. Technical report, RFC 3280, April 2002.
[19] Angela Sasse. Usability and trust in information systems. Technical report, University College London, 2007.
[20] Shunan Ma, Jingsha He, XunboShuai, and Zhao Wang, "Access Control Mechanism Based on Trust Quantification," Social Computing (SocialCom), 2010 IEEE Second International Conference on, Volume:, no., pp: 1032–1037, 20–22 August 2010.

[21] M. Abadi, M. Burrows, B. W. Lampson, and G. D. Plotkin, "A Calculus for Access Control in Distributed Systems," In ACM Trans. Programming Lang. Systems Volume:15, Issue: 4, pp: 706–734, 1993.

[22] Carbone M., Nielsen M., and Sassone V., "A Formal Model for Trust in Dynamic Networks," In International Conference on Software Engineering and Formal Methods, SEFM 2003, IEEE Computer Society, pp: 54–61.

[23] Tommaso Flaminio, G. Michele Pinna, and Elisa B.P. Tiezzi, "A Complete Fuzzy Logical System to deal with Trust Management Systems," In Elsevier journal of Fuzzy Sets and Systems, 2008, Volume: 159, pp: 1191–1207.

[24] S. Weeks, "Understanding Trust Management Systems," In IEEE Symposium on Security and Privacy, pp: 94–105, CA-USA, May 14–16 2001.

[25] R. Mukherjee, B. Baneerjee, and S. Sen, "Learning Mutual Trust," Trust in Cyber-Societies, Berlin, Germany, Springer-Verlag, pp: 145–158, 2001.

[26] R. Falcone, G. Pezzulo, and C. Castelfranchi, "BA Fuzzy Approach to a Belief-based Trust Computation," Lecture Notes in Artificial Intelligence, pp: 73–86, Berlin, Germany, Springer-Verlag, 2003.

[27] K. C. Chen, Y. J. Peng, N. Prasad, Y. C. Liang, and S. Sun, "Cognitive Radio Network Architecture: Part II - Trusted Network Layer Structure," In Proceedings of 2nd International Conference on Ubiquitous Inf. Manage, Communications., pp: 120–124 2008.

[28] Sabrina De Capitani Di Vimercati, Sara Foresti, Sushil Jajodia, Stefano Paraboschi, Giuseppe Psaila, and Pierangela Samarati, "Integrating Trust Management and Access Control in Data-intensive Web Applications," ACM Trans. Web, 6(2):6:1–6:43, June 2012.

[29] A. Jøsang, J. Fabre, J. Hay, J. Dalziel, and S. Pope, "Trust Requirements in Identity Management," In R. Buyya et al., Editors, The Proceedings of the Australasian Informatin Security Workshop (AISW) (Volume 44 of Conferences in Research and Practice in Information Technology), Newcastle, Australia, January 2005.

[30] Fenye Bao, and Ing-Ray Chen, "Trust Management for the Internet of Things and its Application to Service Composition," In IEEE International Symposium on World of Wireless, Mobile and Multimedia Networks (WoWMoM), 2012, Volume:, no., pp: 1–6, June 25–28 2012.

References

[31] Clarke, J., (2008) Future Internet: A Matter of Trust: eMobility Newsletter, http://www.tssg.org/eMobility_Newsletter_200811.pdf accessed 27th July, 2010.

[32] Messaoud Benantar, "Access Control Systems: Security, Identity Management and Trust Models," Springer US, 2006.

[33] Grandison, T, and Sloman, M., Specifying and Analysing Trust for Internet Applications, Proceedings of Second IFIP Conference on e-Commerce, e-Business, e-Government, I3e2002, Lisbon, Portugal, 2002.

[34] Lamsal, P., Understanding Trust and Security, Department of Computer Science, University of Helsinki, Finland, 2001.

[35] Angin, P., Bhargava, B. K., Ranchal, R., Singh, N., Linderman, M., Othmane, L. B., Lilien, L.: An Entity-Centric Approach for Privacy and Identity Management in Cloud Computing. In: IEEE Symposium on Reliable Distributed Systems (SRDS), pp. 177–183 (2010).

[36] Benantar, M., Introduction to the Public Key Infrastructure for the Internet, Prentice Hall, Upper Saddle River, NJ, 2002.

[37] Callas, J., Donnerhacke, L., Finney, H., and Thayer, R., OpenPGP Message Format, IETF RFC 2440, http://www.ietf org, 1998.

[38] L. A. Zadeh, "Fuzzy Sets," In Information and Control Journal, Volume: 8, Issue: 3, pp: 338–353, June 1965.

[39] T. J. Procyk, and E. H. Mamdani, "A Linguistic Self-organizing Process Controller," In Automatica, Volume: 15, pp: 15–30, 1979.

5

Identity Establishment

5.1 Introduction

An identity establishment is nothing but authentication between communicating parties (devices/entities). Authentication of a things does not necessarily mean assurance of identity, but rather assurance of its genuineness or integrity. Authentication is a prerequisite for auditing, accounting and access control, as well as personal application profiles and other services not related to security or accountability [1]. Mobile smart phones, PDA's, palmtops, smart cards, and RFID tags are examples of lightweight devices. There is considerable challenge in balancing, and fine-tuning efficient cryptographic solution on these devices for identity establishment, and access control. Due to diversity of devices, and end users and to protect IoT from well-known attacks, there is a need of attack resistant and lightweight solution for authentication. A security system of IoT can be established by mutual authentication protocol, which can effectively improve the security of communication between the sender and receiver. As key establishment and distribution is fundamental task for entity authentication, it is one of the potential areas to be researched further. In the view of this, it is important to understand the pros and cons of each process. Secret Key Cryptography (SKC) require complicated key pre-distribution, it also require large memory to store key materials and provides low scalability due to key distributions. As compared to SKC, Public Key Cryptography (PUKC) provides more flexible and simple interface. PUKC does not require key pre-distribution, but it requires high energy consumptions and considerable time delay.

Identity establishment is known as the central element to address the security and privacy problems in IoT. It can prevent unauthorized users from getting access to resources, prevent legitimate users from accessing resources in an unauthorized manner, and enable authentic users to access resources in legal manner. This chapter proposes new method of authentication of devices for IoT using public key approach with scalability, and less memory requirements.

5.1.1 Mutual Identity Establishment in IoT

As IoT becomes discretionary part of everyday life, could befall a threat if security is not considered before deployment. The authentication in IoT is important to establish secure communication between devices. Introducing a new device, or user, and achieving authentication and access control to devices/resources in IoT is critical. It is particularly challenging to make authentication and access control secure, efficient, and generic at the same time.

Identity establishment falls under the following three types:

1. Type 1: Something you know (e.g., personal ID number)
2. Type 2: Something you have (e.g., ATM card)
3. Type 3: Something you are (e.g., fingerprints)

Now, in IoT when every possible thing or object that is going to be connected to the Internet, identifying these items/object uniquely has to be done in an efficient and effective manner. The IPV6 addressing mechanism and other related RFID identifiers could be possible ways of providing unique ID's to the objects thereby taking care of the exhaustion of ID's available for identification currently in the IPV4 addressing.

The existence of the devices and their associated sensors can be identified by means of the identification techniques such as IPV6, RFID, barcode and biometrics to identify and authenticate human beings. Tools for managing the identities basically run as applications on dedicated servers or on the cloud. Figure 5.1 explains overall process of mutual identity establishment in general. Sender and receiver send their own credentials to each other and after that one routine is called for credential verifications. Once credentials are verified then mutual identity establishment takes place.

5.1.2 IoT Use Case and Attacks Scenario

Following Figure 5.2, 5.3 and 5.4 presents general use case of attacks possible on IoT. In this use cases, Mobile Entity (x) is a mobile device which represents an entity i.e. any device in the network which communicates with other entities of the same type, or of different types via Internet, or direct. MobileEntity 1, 2, 3 represent three different and most probable scenarios in the system of communication. Possible attacks scenario with the help of use case diagram are described below:

5.1 Introduction

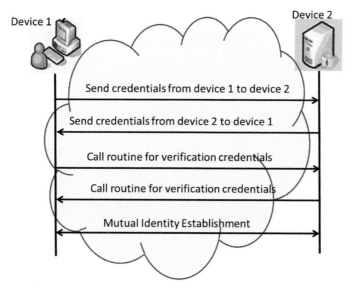

Figure 5.1 General process of mutual identity establishment

i) Denial of Service (DoS) attack

All the devices in IoT have low memory and limited computation resources, thus they are vulnerable to resource enervation attack. DoS attack targets the availability of a system which offers services. This is achieved by consuming resources of the victim so that offered services becomes unavailable to legitimate user. DoS attack is also possible due to man-in-the-middle attack. A general way to launch this attack is to trigger expensive operations at the victim that consume resources, such as bandwidth, computational power, memory or energy. Sample use case of DoS in IoT is shown in Figure 5.2. There are multiple mobile entities communicating with each other and because of malicious attacker response of one entity to another entity is being delayed since there is DoS attack. Communication can be through private network, or it can be through remote communication, or it can be through secure channel, DoS attack makes the service unavailable to legitimate user. Sometime smart devices accessing some channels and those channels itself are flooded because of DoS attack.

ii) Man-in-the-middle attack

Man-in-the-middle attack is one type of eavesdropping possible in the commissioning phase of devices to IoT. Man-in-the-middle attack takes place when malicious entity is on the network path of two legitimate entities.

106 Identity Establishment

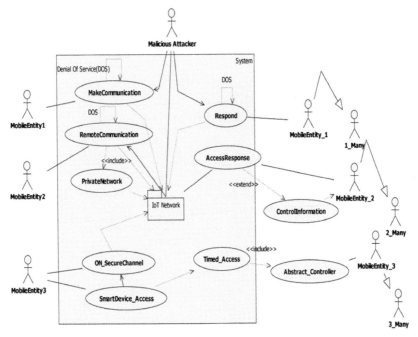

Figure 5.2 Use case for denial of service attack

The key establishment protocol is vulnerable to man-in-the-middle attack and can compromise device authentication as devices usually do not have prior knowledge about each other. Hence it is capable of delaying, modifying message. Sample use case of man-in-the-middle attack in IoT is shown in Figure 5.3. As stated earlier there are multiple mobile entities communicating with each other. There can be man-in-the-middle which can be placed on the network path of two different mobile entities. This malicious attacker is unauthorised listener who does not attempt to break the keys of involved entities but rather to become falsely trusted man-in-the-middle. Unauthorized user achieves this by replacing the exchanged keys with its own. In this way each of entities establish secure channel with malicious users.

iii) Replay attack

As discussed in chapter 2, during the exchange of identity related information or other credentials in IoT, this information can be spoofed, altered or replayed to repel network traffic. This causes a very serious replay attack. For example, during the authentication process of two devices, an attacker can collect authentication information from both the devices and can use this information

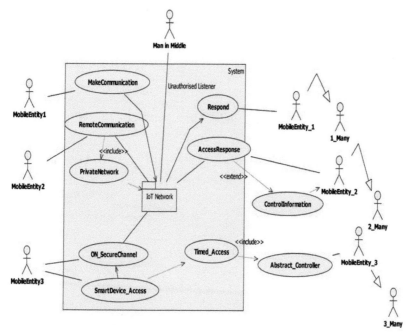

Figure 5.3 Use case for man-in-the-middle attack

in future for personal use. This information can also be used by an attacker to delay or repeat the transmission and referred as replay attack. Replay attack can also cause DoS attack acting as legitimate entity. Replay attack is essentially one form of active man-in-the-middle attack. Sample use case of replay attack in IoT is shown in Figure 5.4. Authorised user's login information may be captured by replay attacker and resent this information again to get access of services from other mobile entities.

5.1.3 State of the Art

There is a large research done in the area of securing IoT. There is a closely related work done in the MAGNET project [2, 3] where security association takes place with the increased communication overhead, and authentication is left unaddressed. The authors presented distributed access control solution based on security profiles, but attack resistance is not explored. In [4, 5], the authors have presented ECC-based authentication protocol, but the major disadvantage is that it is not DoS attack resistant. As in IoT, there are billions of devices, and resistance to DoS attack is one of the most important issues.

108 Identity Establishment

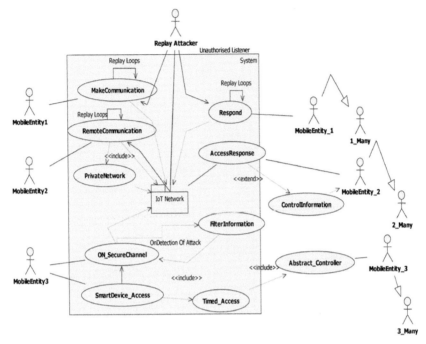

Figure 5.4 Use case for replay attack

In [6], the author addresses the problem of secure communication, and authentication based on shared key, and is applicable to limited location, and cannot be used for a wide area. It addresses the peer-to-peer authentication but cannot be extended in resource constrained environment.

In [7], the author has presented ECC-based mutual authentication protocol for IoT using hash functions. A mutual authentication is achieved between terminal node, and platform using secret key cryptosystem introducing the problem of key management, and storage. Self-certified keys cryptosystem based distributed user authentication scheme for WSN is presented in [8] where only user nodes are authenticated, and is not a lightweight solution for IoT. In [9], the author presents authentication with parameter passing during the handshake. Handshake process is time consuming, and based on symmetric key cryptography with more memory requirement for large prime numbers. Efficient identification and authentication is presented in [10], which is based on the signal properties of node but is not suited for mobile nodes. Direction of the signal is considered as a parameter for node authentication, but it takes more time to decide signal direction with more memory, and computations involved. In [11], a cluster-based authentication is proposed which is most suited for futuristic IoT, but an attacker can get hold of the distribution of system key pairs, and cluster key. Generation of random numbers and signatures creates considerable computational overhead consuming memory resources.

State of the art evaluation is shown in Table 5.1. The related work is summarized based on the parameters like mutual authentication, lightweight solution, resistant to attacks, distributed nature, and access control solution. Recent related work in the area of authentication for IoT is considered for the evaluation, and is presented below. From Table 5.1, it is clear that, all existing solutions for authentication and access control do not fulfill all requirements for IoT. Blue block in the Table 5.1 represents respective feature unavailability in corresponding solutions.

5.2 Cryptosystem

Cryptography word comes from *Greek* Dictionary. Word *kryptós* means "hidden" and *gráphein* means "to write". In short cryptography is art and science of writing secrets by encoding messages and makes them non-readable. Study of many encryption principles/methods is also known as cryptography. Use of such principle or method in system is known as cryptosystem. There has been lot of debate about which of the cryptographic primitives like public key, or private key is suitable for IoT. Most of the research has mainly focused in the area like Wireless Sensor Network (WSN), and its application like healthcare, and smart home. Many security mechanisms have been proposed based on private key cryptographic primitives due to fast computation, and energy efficiency. The scalability problem and memory requirement to store keys makes it inefficient to heterogeneous devices in IoT. Public key cryptography-based solution overcomes these challenges with high scalability, low memory requirements and no requirement of key pre-distribution infrastructure.

There are two important aspects to secure the communication. One is the confidentiality and other is authenticity. If you want to send some email which contains some important or sensible information, you have to assure that information must not be accessible to anybody else. Confidentiality can be achieved by private key or public key cryptography and choice depends on the underlined application. If you want to access bank site through browser and when you enter web address for that, you get back the website for your bank. You want to be assure that this is your bank website and not malicious site that attempting to compromise your account. Public key cryptography solve this problem to some extends.

110 *Identity Establishment*

Table 5.1 State of the art evaluation summary

Parameters	Mutual Authentication	Lightweight Solution	Attack Resistant				Distributed Nature	Access Control
Solutions			DoS	Man in Middle	Replay			
[2, 3]	×	×	×	×	×		✓	✓
[4, 5]	✓	✓	×	×	×		✓	×
[6]	×	×	✓	✓	✓		×	×
[7]	✓	✓	×	✓	✓		✓	×
[8]	×	×	×	×	×		✓	×
[9]	✓	✓	×	✓	✓		×	×
[10]	✓	×	×	×	×		✓	×
[11]	✓	×	×	×	×		✓	×

[2, 3]: Ubiquitous access control in MAGNET
[4,5]: ECC based Authentication in RFID
[6]: Authentication in Ad-hoc Wireless N/W
[7]: Authentication in IoT
[8]: Authentication in WSN
[9]: Progressive Authentication in Ad-hoc N/W
[10]: Peer Identification and Authentication
[11]: Authentication in Ad-hoc N/W

5.2.1 Private Key Cryptography

Private key cryptography is conventional cryptography which is also known as symmetric key encryption or secret key encryption.

As shown in Figure 5.5 secrete key is used to encrypt the data and same key is used to decrypt the data. This mechanism of private key cryptography makes the computations faster as compared to public key cryptography. The problem with this encryption is, in order to decrypt the information the key must be available. This causes following two problems:

1. Key must be stored securely and only accessible when require. If attacker will get access to those keys he can decrypt those messages using the same key which is used to encrypt the message.
2. Other parties want to decrypt information using the same key, that time secure channel is require to transfer the key.

5.2.2 Public Key Cryptography

Public key cryptography is known as asymmetric key encryption. As shown in Figure 5.6, in public key cryptography, with the help of strong mathematics base we can generate pair of keys which have unique relation. One of the key is used to encrypt the information which is known as public key and the other associate key used to decrypt the information which is known as private key. Public key cryptography represents an effective approach to data encryption as it can provide an increased level of confidence for exchanging

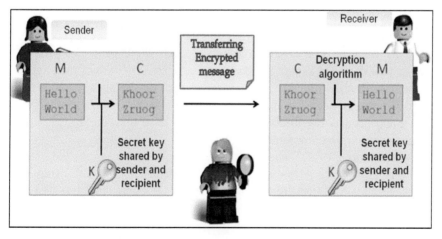

Figure 5.5 High level view of private key cryptography

112 Identity Establishment

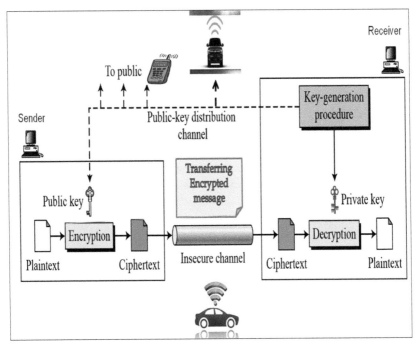

Figure 5.6 High level view of of public key cryptography

information over an increasingly insecure environment. In case the public key gets compromised, still it is not computationally feasible to retrieve the private key.

Public key cryptography is also used in the process of authentication. Sender uses private key of its own to sign message and public key of receiver to encrypt the message. This encrypted message sends to other recipient. Now recipient can make use of sender public key to validate signature of message. So receiver will highly confident that owner of the private key was the one who creates message. In this way, we can use combination of public key for encryption and private key for signing in order to have both confidentiality and authenticity for the communication.

Consider example of IoT and healthcare, devices that generate patient-related information for e.g. body sensor readings can encrypt data using a public key and the health monitoring applications (e.g., cloud or web systems operated by relatives) can use the private key to decrypt the data. Using public key cryptography digital certificates the proper authentication of the devices can be achieved, in addition to the secure data transmission.

5.3 Mutual Identity Establishment Phases

This chapter presents mutual identity establishment scheme for IoT. We recommend that, for any identity establishment scheme, following phases have to be considered in general:

1. Secret key generation
2. One way authentication
3. Mutual authentication

Details of each phase are described in as follows.

5.3.1 Secret Key Generation

There is considerable interest in Elliptical Curve Cryptography (ECC) for IoT security [12]. It has advantages of a small key size, and a low computation overhead. It uses a public key cryptography approach based on elliptic curve on finite fields. Elliptical Curve Cryptography-Diffie Hellman (ECCDH) [12] is a symmetric key agreement protocol that allows two devices that have no prior knowledge about each other to establish a shared secret key which can be used in any security algorithm. Using this public parameter, and own private parameter, these parties can calculate the shared secret. Any third party, who doesn't have access to the private details of each device, cannot calculate the shared secret from available public information. All the devices joining IoT share key pairs during the bootstrapping. Mutual Identity Establishment and Capability-based Access Control (MIECAC) scheme presented in this chapter is also applicable to security bootstrapping as well as access control. The security bootstrapping is the process by which devices join IoT with respect to location, and time. It includes device authentication along with credential transfer. The protocol uses one, or more trusted Key Distribution Center (KDC) to generate domain parameter and other security material, and important part is this KDC is not required to be online always. Initially KDC randomly selects particular elliptic curve over finite field $GF(p)$ where p is a prime, and makes base point P with large order q (where q is also prime). KDC then picks random $x \varepsilon\ GF(p)$ as a private key, and publishes corresponding public key $Q = x \times P$. KDC generates random number $K_i \varepsilon\ GF(p)$ as a private key for device i and generates corresponding public key $Q_i = K_i \times P$. The key pair $\{Q_i, K_i\}$ is given to device i. With an increasing number of devices, the KDC can generate ECC key pair based on base point P for any number of devices as it is rich in terms of resources as compared to other devices in IoT. These ECC key pairs will be used to share a common secret key for secure communication using ECCDH,

114 *Identity Establishment*

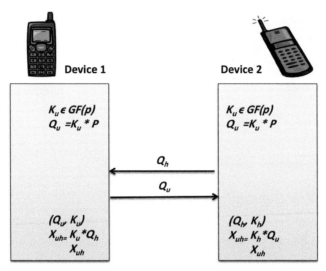

Figure 5.7 ECCDH for establishing shared secret key

and is explained below. Steps of aforementioned ECCDH are presented in Figure 5.7.

Assumption is that ECC is running at trusted KDC. There is an agreement on system based point P and generate (Q_u, K_u) and (Q_h, K_h) pairs where

Q_u = Public key of Device 1
K_u = Secret key of Device 1
Q_h = Public key of Device 2
K_h = Secret key of Device 2

And *P* is large prime number over *GF (P)* and generations of above keys are as follows:

No parameter is disclosed in this process of establishing a shared secret key other than domain parameter *P*, and public keys. This chapter considers sensor node as device, because the functionalities and operational principle of wireless sensor networks makes it appropriate, and mandatory candidate of IoT.

5.3.2 One Way Authentication

One way authentication authenticates Device 1 to Device 2, and is explained below. As per above ECCDH, both Device 1 and Device 2 has X_{uh} as a common secret key. Device 1 selects r ∈ GF (P) which will be used to create session key. T_u is generated as a time stamp by Device 1. It is assumed that synchronisation has taken care by using a appropriate mechanism. The secret key is created by Device 1 as $L = h(X_{uh} \oplus T_u)$. Then, Device 1 encrypts r with secret key L as $R = E_L(r)$ and encrypts T_u by X_{uh} as $T_{us} = E_{X_{uh}}(T_u)$. After this Device 1 builds a Message Authentication Code (MAC) value as $MAC_1 = MAC(X_{uh}, R \| ICAP_1)$ where $ICAP_1$ is a data structure representing an Identity-based Capability for this Device 1 giving access rights. Details about ICAP are given in the Chapter 6 in detail. Now Device 1 sends following parameters to Device 2 directly, or through gateway node / coordination node, or access point as (R, T_{us}, MAC_1). Device 2 generates it's current time stamp as $T_{current}$, and Device 2 will decrypt T_{us} to get T_u and compare it with $T_{current}$. If $T_{current} > T_u$, it is valid.

Now Device 2 calculates *L,* and decrypt *R* to get *r*. Device 2 also calculates the *MAC_1'*, and it will verify this with *MAC_1* received from Device 1. If valid, then Device 1 is authentic to Device 2. Device 1 also matches the *$ICAP_1$* received with *$ICAP_2$* stored at Device 2. If Device 2 gets match with *R, MAC_1,* T_{us} then Device 1 is authenticated to Device 2. Aforementioned protocol is presented in Figure 5.8.

5.3.3 Mutual Authentication

This part of authentication authenticates Device 2 to Device 1, and is explained below in Figure 5.9. Device 2 builds a MAC as $MAC_2 = MAC(r \| ICAP_2)$ and also encrypts r with X_{uh} as $R' = E_{X_{uh}}(r)$. Device 2 sends (R', MAC_2) to Device 1. Device 1 verifies MAC_2, and decrypt R' and compare the received r with this r (denoted as r' and r" in figure). If a match is found, Device 2 is also authenticated to Device 1, and communication, and access will be granted based on the $ICAP_2$. This protocol achieves both mutual authentication along with capability-based access control in secure way.

5.4 Comparative Discussion

As security protocols are event driven and sensitive in nature, emphasis should be given on integrating formal verification of security protocol in design, and development phase. Any security protocol should take into account mainly two design goals: reduce the overhead that protocol imposes on underlying resource constrained environment, and provide reasonable protection for security attributes that are targeted. So it is essential to verify new

116 Identity Establishment

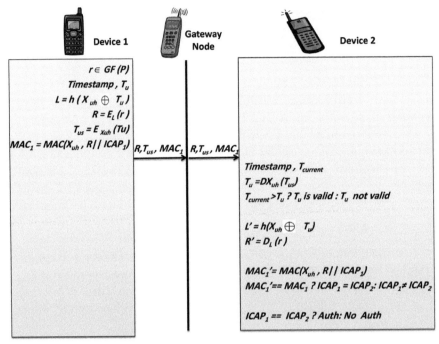

Figure 5.8 One way authentication protocol

implemented security protocol by globally accepted automatic tool based on formal specification language for protocol input.

5.4.1 Security Protocol Verification Tools

Automated Validation of Internet Security Protocols and Applications (AVISPA) [14] provide formal language as High Level Protocol Specification Language (HLPSL) to input new authentication protocol, and validate them. The AVISPA project aims at developing a push-button, industrial-strength technology for the analysis of large-scale Internet security-sensitive protocols and applications. HLPSL is a language developed by the AVISPA IST project. It is partially based on temporal logic of actions and is explicitly designed to validate security protocols. Actions and events can be implemented using HLPSL. AVISPA works in stages. Protocol model is specified using HLPSL; it is translated into Intermediate Format (IF). IF is mapped by input of many backend: SATMC (SAT base Model Checker), OFMC (On the Fly Model Checker), Cl-Atse (Constraint-Logic-based Attack searcher), and TA4SP

5.4 Comparative Discussion

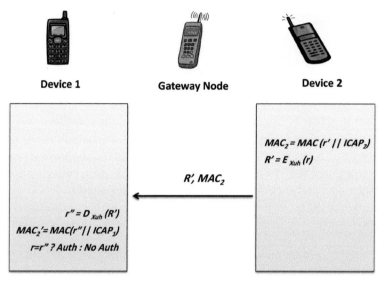

Figure 5.9 Protocol for mutual authentication

(Tree Automata Based on Automatic Approximation for the Analysis of Security Protocols). The AVISPA framework is not the only tool available for security protocol verification. There have been several other efforts in this area. ProVerif [15] which is based on Horn Clauses and Pi-calculus, Scyther [16] based on symbolic backward search, Casper/FDR [17] based on process algebra, Symbolic Trace Analyzer (STA) [18] based on variant of Pi-calculus and the Brurus [19] which is logic based are few notable efforts in the area of security protocol verification tool.

The few reasons why the AVISPA is useful tool are as follows:

- The AVISPA tools set is outcome of recent effort with developed set of tools and methods.
- The AVISAP is actively maintained by active user community.
- Integrates different back-ends implementing a variety of state-of-the-art automatic analysis techniques for
  - Protocol falsification (by ending an attack on the input protocol)
  - Abstraction-based verification methods both for finite and infinite numbers of sessions

Security in IoT is critical due to the dynamic network topology, and nomadic nature. An intruder can intercept messages, cause replay attacks, steal identity, or inject false messages. Such kind of intruders are presented in [20],

118   *Identity Establishment*

and known as Dolev-Yao intruders. AVISPA uses Doley-Yao intruder model which is more suitable for IoT, and is the strongest model. Many researchers have analyzed security protocol [21] for WSN using AVISA and reported security flaws if any. AVISPA is also used by Internet Engineering Task Force (IETF) and International Telecommunications Union (ITU). Protocol is written in CAS+ format and then using AVISPA tool, it is converted into HLPSL, and then it is simulated with AVISPA. ATSE, and Verbose test for proposed protocol using Doley-Yao intruder model shows that protocol is not prone to attacks. Size, number of messages to reduce memory requirement and bandwidth usage are the main performance parameters of security protocol for IoT. Efficiency, and security design of protocol, presented in this chapter is validated by AVISPA.

Implementation of protocol can be as follows:

1. First stage of protocol authenticates Device 1 to Device 2, and i.e. one way authentication,
2. Second stage of protocol is for mutual authentication i.e. authenticates Device 2 to Device 1.
3. Every entity *(Device 1, Device 2, Gateway_Node)* is translated into HLPSL agent code to specify action, and sessions are built.
4. An important point arises that an intruder can impersonate any agent by putting fake variable instead of agent, and can receive all messages. AVISPA uses channel (dy) which assumes that intruder can intercept every message in the channel, and can create any message from the intercepted message. However, this model works on the principle of perfect cryptography which implies that the intruder cannot decrypt messages encrypted with key k with another key k' different from k.
5. AVISPA provisions to create environment with sessions, intruder knowledge and attack to be targeted. The intruder knowledge includes security parameters, identities etc.

5.4.2 Security Analysis

The attacks covered in this chapter are man-in-the-middle attack, replay attack *(authentication_on, request, witness),* and DoS. An incremental methodology with multi-session is used to validate the protocol by increasing knowledge to intruder except shared secret key. The evaluation will focus on identity establishment in terms of one way, and mutual as the most important processes in the authentication. Verification results are described below.

A. Evaluation procedure

In order to carry out the evaluation using AVISPA some assumptions can be made. Both the devices have already obtained ECC-based shared key using Diffie-Hellman (ECCDH). As stated earlier, the assumption here is that KDC is secure, and trusted. Complete protocol evaluation is presented in following model:

$$D_1 \rightarrow D_2 : [R, T_{us}, MAC_1]; [\{r\}\_L, \{T_u\}\_X_{uh}, RND_1]$$
$$D_1 \leftarrow D_2 : [R', MAC_2]; [\{r\}\_X_{uh}, RND_2]$$

Where:

- D_1: Device 1
- D_2: Device 2
- $\{\ \}\_$: A symbol of encryption
- T_u: Timestamp generated as a nonce
- X_{uh}: A shared key between $D1$ and $D2$ using ECCDH
- r : Some value $x \in GF(p)$
- RND_1 : MAC value of X_{uh}, R and $ICAP_1$ where $ICAP$ is result of one way hash function $f(Device\_ID, Access\ Rights, Rnd)$, Rnd is random number generated to prevent forgery
- RND_2: MAC value of r and $ICAP_2$
- L : result of one way hash function (XOR of X_{uh} and T_u)

Besides this, Dolev-Yao intruder model has been introduced in the evaluation. The intruder is assumed to have the knowledge of the following:

- ID: Device identifier
- $f()$: Knowledge of one way hash function

B. Evaluation results

The goal of evaluation is to verify protocol for attacks mentioned above, and ensure mutual authentication along with access control.

- **Mutual authentication**

X_{uh} is shared securely between D_1 and D_2 and r is provided by trusted KDC to both the devices. Consequently, D_1 is authenticated to D_2 as only D_2 can decrypt R and T_{us}. Also MAC can be calculated only by D_2 and D_2 is sending encrypted r to authenticate it to D_1. Verification results show that secure mutual authentication is achieved.

- **Man-in-the-middle attack**

In case of authentication, even there is man-in-the-middle attack on R, T_{us}, MAC_1 parameters; attacker will not reveal any information. AVISPA shows that authentication protocol is free from this attack. For access control, man-in-the-middle attacks happen when an attacker eavesdrop the *ID*, and *ICAP* transmitted, and then masquerade attacks happens when the attacker uses the stolen *ID*, and *ICAP*. The key to preventing masquerade attack from the stolen *ICAP* is to use *ID* to validate the correct device. If the attacker manages to steal the *ID*, the attack is prevented by applying public key cryptography to *ID*, assuming that the authentication process has been done before access control. In this way, although the attacker gets the *ICAP* which is not encrypted, the capability validity check will return an exception because of the one way hash function, *f(ID, AR, Rnd)* will return a different result than the one presented in the *CAP*, without a correct *ID*.

Another type of man-in-the-middle attack is replay attack. Adversary can intercept the message sent out from D_1. However, it is not possible in MIECAC because it can easily detect by verifying timestamp T_u. If T_u is older than predefined threshold value, it is invalid, and has been used. If T_u is changed, $MAC_1 = MAC(X_{uh}, R \,||\, ICAP_1)$ is not valid and consistent. For access control, MIECAC prevents the replay attack by maintaining the freshness of Rnd, for example by using time stamp, or nonce by including *MAC* as well. Even if the attacker manages to compromise the solution and gets the *ICAP*, it cannot use the same capability the next time because the validity will be expired.

- **DoS attack**

Upon receiving the message from D_1, D_2 first check the validity of timestamp. If it is not valid, then D_2 discards the message. Otherwise, it computes a MAC_2 value to compare with received value. DoS happens when an attacker accesses a particular resource massively, and simultaneously by using the same, or different *ID*s. It is easy to control access using one *ID* because the system is able to maintain the session, thus the access of the same *ID* to the same resource can be restricted to only one session at a time. The potential of DoS attacks from multiple *ID* s can be prevented in the capability propagation process. Therefore, DoS attack can be prevented, or at least minimized.

As authentication refers to identity establishment and identity privacy refers to the problem of ensuring that communication takes place only between right devices without disclosure of identity information to unauthorized

5.4 Comparative Discussion

eavesdroppers. Eavesdropping is another threat in the absence of authentication. When two devices are communicating to each other, third device in between these two devices can listen to entire communication and get the authentication information. During the authentication process of two devices, an attacker can collect authentication information from both the devices and can use this information in future for personal use thus violating identity privacy. This scenario is depicted in the Figure 5.10. Example scenario is: The WSN deployed for the homeland security scenario may include features (e.g., sensors equipped with cameras) that allow the tracking of the movement of suspicious individuals in a given area. An attacker may try to misuse these features by tracking the whereabouts of innocent people.

There are many use case scenarios of IoT like agriculture, smart home and land sliding. The events which endanger IoT from security and privacy point of view are threats to IoT. Threat analysis for IoT in this chapter is done by defining negative scenarios referred as misuse case. The main assets in IoT are resources (data) and devices in the context of underlined scenario. By analyzing many scenarios, this chapter proposes following 4

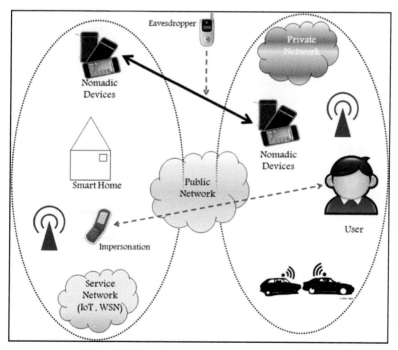

Figure 5.10 Attack on identity/location privacy

122 Identity Establishment

general objectives of an attacker as follows with respect to the adversary model presented in threat modelling section of chapter 2:

- Illegitimate access to the information/resources provided by IoT,
- Falsification of information provided by IoT,
- Denial of service i.e. disturbing the operation of IoT fully or partially,
- Movement and action tracking of individuals or devices

Hiding device idetifiers and location identifiers from neighbouring as well as intermediate devices is necessary to achieve identity/location privacy. Ensuring privacy is equivalent to ensuring that there is no man-in-the-middle attack for communication between two devices. In the proposed MIECAC scheme, it is seen that even there is man-in-middle attack on R, T_{us}, MAC_1 parameters; attacker will not reveal any information. When two devices communicating to each other exchange localization and tracking details with each other, mutual authentication as presented as MIECAC scheme in this contribution will ensure location privacy in the absence of man-in-the-middle attack.

5.4.3 Performance Metrics

Security level of protocol presented in this chapter depends on the type of MAC algorithm, encryption algorithm, and security level of ECC signature. We propose to use RC5 stream cipher for encryption, which takes 0.26 ms on Mica2 motes [22–24]. RC5 is notable for its simplicity for resource constrained devices such as IoT and its flexibility due to the built in variability. Heavy use of data independent rotations, and mixture of different operations provides strong security to RC5 [25].

We can use SHA-1 as one way hash function which takes 3.63 ms on Mica2 motes, and it is computationally expensive to find text which matches given hash, and also it is difficult for two different texts which produces the same hash [22–24]. To generate the MAC value, we propose CBC-MAC which has advantage of small key size and small number of block cipher invocations and takes 3.12 ms on Mica2 motes [23]. The time required to generate random number is 0.44 ms, and ECC to perform point multiplication which takes 800 ms on Mica2 motes [23, 24]. In MIECAC protocol as the message length is fixed, CBC-MAC is most secure [26]. It is clear from these values that maximum time is required for ECC point multiplication. In MIECAC, point multiplication is taking place at KDC, and as KDC is a powerful device, computational overhead is trivial as compared to the sensors. We denote the computational time required for each operation by device in IoT by the following notation:

D_H = Time to perform one way hash function SHA-1

D_{MAC} = Time to generate MAC value by CBC-MAC

D_{RC5} = Time to perform encryption and decryption by RC5

D_{MUL} = Time to perform ECC point multiplication

R = Time for random number generation

Table 5.2 shows the comparison of computational time for the above-mentioned protocol. MIECAC protocol for mutual authentication and access control for IoT devices takes less time (14.02 ms) as compared to other protocol compared in this chapter. Key point to note here is that, none of the work has addressed the issue of authentication, and access control as an integrated solution for IoT. Total computational time for of the proposed scheme, HBQ [16], and mutual authentication for IoT (IoT_Auth) [7] is shown in Table 5.2. IoT_Auth scheme requires $R + D_H + 2D_{MUL}$ time for mutual authentication which comes approximately 1604.07 ms. HBQ scheme takes $2D_H + 2D_{MAC} + D_{RC5} + 3D_{MUL}$ total time for authentication which is approximately 2,413.76 ms. Key point to note here is that both the schemes do not address access control after authentication. MIECAC takes only $D_H + 2D_{MAC} + 2D_{RC5}$ which takes only 14.02 ms which is much better than other two schemes analyzed in this chapter. In MIECAC, $2D_H$ factor is introduced which comprises time required by one way hash function in authentication as well as in *ICAP* to calculate *Rnd*. Due to unbounded number of devices in IoT, each device should not authenticate in short time due to unbounded number of devices, and receipt of their authentication request at the same time. Therefore, secure, and efficient group authentication, and authorization scheme is required that authenticates a group of devices at once in the context of resource constrained IoT. Threshold Cryptography-based Group Authentication (TCGA) [28] scheme for IoT which verifies authenticity of all the devices taking part in the group communication is promising and efficient approach.

Table 5.2 Computational time for MIECAC [13]

Scheme	Auth. Time	Total time
IoT_Auth [7]	$R + D_H + 2D_{MUL}$	1604.07 ms
HBQ [27]	$2D_H + 2D_{MAC} + D_{RC5} + 3D_{MUL}$	2413.76 ms
MIECAC	$2D_H + 2D_{MAC} + 2D_{RC5}$	14.02 ms

5.5 Conclusions

A distributed, lightweight, and attack resistant solution, being the most favorable choices for IoT, puts resilient challenges for authentication and access control of devices. This chapter has discussed general identity establishment process and essentially in the context of IoT, how attack happens. IoT use case is presented and behavioral modeling of the attacks like man-in-the-middle, replay and DoS attack is presented and discussed using unified modeling language-based use case approach. The comparison of public and private key cryptographic primitives and their application to the IoT security is also discussed. The detailed survey of different authentication scheme as well as evaluation of these scheme based on few performance parameters is discussed to understand the gap analysis in next part of this chapter.

In the sequel, efficient, and scalable ECC-based authentication, and access control protocol is presented in next part of this chapter. Protocol is divided in two phases as one way authentication, mutual authentication, and integrated with capability-based access control solution. Power of ECC is extended to achieve mutual authentication of devices with novel capability-based approach for access control. Furthermore, this chapter presents comparative analysis of different authentication, and access control schemes for IoT. A comparison in terms of computational time shows that MIECAC scheme is efficient as compared to other solutions. Protocol is also analyzed for the performance, and security point of view for different possible attacks in IoT scenario. Protocol evaluation shows that it can defy attacks like DoS, man-in-the-middle, and replay attacks efficiently, and effectively. This chapter also presents protocol verification using AVISPA tool which proves that the MIECAC protocol is also efficient for large scale devices in terms of key sharing, and authentication.

References

[1] Anders Fongen, "Identity Management and Integrity Protection in the Internet of Things," est, pp.111–114, 2012 Third International Conference on Emerging Security Technologies, 2012.

[2] Ramjee Prasad, "My personal Adaptive Global NET (MAGNET)," Signals and Communication Technology Book, Springer Netherlands, Pages: 435, 2010.

[3] Kyriazanos Dimitris M., Stassinopoulos George I., and Neeli R Prasad, "Ubiquitous Access Control and Policy Management in Personal Networks," In Third Annual IEEE International Conference on Mobile and Ubiquitous Systems: Networking & Services, Volume: Issue: pp:1–6, San Jose-CA July 17–21 2006.

[4] Michael Braun, Erwin Hess, and Bernd Meyer, "Using Elliptic Curves on RFID Tags," In IJCSNS International Journal of Computer Science and Network Security, Volume: 8, Issue: 2, pp: 1–9 2008.

[5] Sheikh Iqbal Ahamed, Farzana Rahman, and Endadul Hoque, "ERAP: ECC based RFID Authentication Protocol," In 12th IEEE International Workshop on Future Trends of Distributed Computing Systems, pp: 219–225. Kunming, October 21–23 2008.

[6] Balfanz, D., Smetters D. K., Stewart P., and Wong H. C., "Talking to Strangers: Authentication in Ad-hoc Wireless Networks," In Network and Distributed System Security Symposium; pp: 6–8, San Diego CA-USA, February 2002.

[7] Guanglei Zhao, Xianping Si, Jingcheng Wang, Xiao Long, and Ting Hu, "A Novel Mutual Authentication Scheme for Internet of Things," In Proceedings of 2011 IEEE International Conference on Modeling, Identification and Control (ICMIC), Volume: Issue: pp: 563–566, Shanghai – China, June 26–29 2011.

[8] C. Jiang, B. Li and H. Xu, "An Efficient Scheme for User Authentication in Wireless Sensor Networks," In 21st International Conference on Advanced Information Networking and Applications Workshops, pp: 438–442, Niagara Falls – Ont, May 21–23 2007.

[9] R. R. S. Verma, D. O'Mahony, and H. Tewari, "Progressive Authentication in Ad-hoc Networks," In Proceedings of the Fifth European Wireless Conference, Barcelona – Spain, February 24–27 2004.

[10] Suen, T., and Yasinsac A., "Ad-hoc Network Security: Peer Identification and Authentication using Signal Properties," In Proceedings from the Sixth Annual IEEE SMC Information Assurance Workshop, IAW '05, Volume:, no., pp: 432–433, NY-USA, June 15–17 2005.

[11] Venkatraman L., and Agrawal, D.P., "A Novel Authentication Scheme for Ad-hoc Networks," In IEEE Wireless Communications and Networking Conference, WCNC-2000, Volume: 3, no., pp:1268–1273, Chicago-IL, 2000.

[12] N. Koblitz, "Elliptic curve cryptosystems," in Mathematics of Computation, Volume: 48, pp: 203–209, 1987.

[13] Parikshit N. Mahalle, Bayu Anggorojati, Neeli R. Prasad, and Ramjee Prasad, "Identity Establishment and Capability Based Access Control (IECAC) Scheme for Internet of Things," In proceedings of IEEE 15th International Symposium on Wireless Personal Multimedia Communications (WPMC – 2012), pp: 184–188. Taipei - Taiwan, September 24–27 2012.

[14] Avispa – A tool for Automated Validation of Internet Security Protocols. http://www.avispa-project.org.

[15] Blanchet, B., "An Efficient Cryptographic Protocol Verifier based on Prolog Rules," In Proceedings of the 14th IEEE workshop on Computer Security Foundations (Washington, DC, USA, 2001), CSFW '01, IEEE Computer Society, pp. 82–96.

[16] C. Cremers. Scyther, "Semantics and Verification of Security Protocols," Ph.D. Dissertation, Eindhoven University of Technology, 2006.

[17] G. Lowe. Casper: a compiler for the analysis of security protocols. J. Computer. Security, (1–2): 53–84, 1998.

[18] Michele Boreale and Maria Grazia Buscemi, "Experimenting with STA, a tool for automatic analysis of security protocols," In Proceedings of the 2002 ACM symposium on Applied Computing, Madrid, Spain, pages 281–285, ACM Press, 03 2002.

[19] E. M. Clarke, S. Jha, and W. Marrero, "Verifying security protocols with Brutus," In ACM Transactions on Software Engineering and Methodology (TOSEM), 9(4): 443–487, 10 2000.

[20] D. Dolev and A. C.-C. Yao, "On the Security of Public Key Protocols," In IEEE FOCS, pp: 350–357, 1981.

[21] Mihai Lica Pura, Victor Valeriu Patriciu, and Ion Bica, "Formal Verification of Secure Ad-hoc Routing Protocols Using AVISPA: ARAN Case Study," In ACM Proceeding ECC'10 Proceedings of the 4th conference on European computing conference 2010, pp: 200–206, Bucharest – Romania, April 20–22 2010.

[22] R. Chakravorty, "A Programmable Service Architecture for Mobile Medical Care," 4th IEEE International Conference on Pervasive Computing and Communications, 2006, pp: 36–55, Pisa, March 13–17 2006.

[23] C. Karlof N. Sastry, and D. Wagner, "Tinysec: Link Layer Security Architecture for Wireless Sensor Networks," In SensSys, ACM Conference on Embedded Networked Sensor Systems, 2004, pp: 162–175, Baltimore – MD – USA, November 3–5 2004.

[24] N. Gura A. Patel, A. Wander, H. Eberle, and S. C. Shantz, "Comparing Elliptic Curve Cryptography and RSA on 8-it CPUs," In Proceedings of Cryptographic Hardware and Embedded Systems 2004, Volume: 3156, LNCS, pp: 119–132, Cambridge MA-USA, August 11–13 2004.
[25] Y. L. Yin, "The RC5 Encryption Algorithm: Two Years On," CryptoBytes (3) 2 (Winter 1997).
[26] M. Bellare J. Killan, and P. Rogaway, "The Security of Cipher Block Chaining," CRYPTO '94 Proceedings of the 14th Annual International Cryptology Conference on Advances in Cryptology, LNCS, Volume: 839, pp: 341–358. Springer, Heidel-verg 1994.
[27] H. Wang B. Sheng, and Q. Li, "Elliptic Curve Cryptography based Access Control in Sensor Networks," International Journal of Security and Networks, Volume:1, Issues: 3/4, pp. 127–137 2006.
[28] Parikshit N. Mahalle, Neeli R. Prasad and Ramjee Prasad, "Novel Threshold Cryptography-based Group Authentication (TCGA) Scheme for the Internet of Things (IoT)," In proceedings of IEEE 4th International Conference on Wireless Communications, Vehicular Technology, Information Theory and Aerospace & Electronic Systems Technology (Wireless ViTAE – 2014). Aalborg – Denmark, May 11–14 2013.

6
Access Control

6.1 Introduction

Due to unbound number of things which includes resources, devices, and services, IoT has a more demanding and challenging environment in terms of scalability, and manageability. In IoT, users, and devices are able to create profiles and according to the situation, and the context, the access is granted to the resources. These ideas are very well documented in the available literature. Representative examples are "Scenarios for Ambient Intelligence" in 2010 [1] and, the vision of Association of Computing Machinery in "The next 1000 Years [2]. MAGNET [3] is another example of IoT application which is an integrated project supported within the Sixth Framework Programme (FP6) of the European Union (EU) commission. The project gives full emphasis on personalization, access control, and personal networking. These scenarios envisage that IoT specific approaches are distributed, and ad-hoc in nature. With dynamic network topology, management of IoT networks become lurid if the management of authorization and access control is not addressed. The devices ranging from sensors to RFID tags, identities extended to devices, ubiquitous interaction, and large numbers of heterogeneous devices are the main challenges of IoT to design security solutions. Access control and authorization in IoT with the least privilege is equally important to establish secure communication between multiple devices, and services. The requesting entity is referred to as the SUBJECT, and the entity to be accessed is referred to as the OBJECT in access control terminology. In IoT context, there are many subjects that need to access resources, for example: preventive smart home maintenance, and ubiquitous health care applications. The access control is also critical due to its potential impact on the behavior of the system, but also there is an access to sensitive information, or services that are available. The principle of the least privilege is an important feature of access control solution which limits the access to minimum resources which are required,

130 *Access Control*

and also referred to as selective access. In the context of IoT, the principle of least privilege is preferred.

Different applications of IoT are presented and discussed in Chapter 1. The main challenges in these application areas are to ensure that ubiquitous access to services and monitoring data is granted to identities that fulfill the access control rules for IdM, heterogeneous device interaction, authorization, mutual authentication and secure delegation from a mobile device, and the secure data access. Securing user interactions with IoT is essential if the notion of "things everywhere" is to succeed. In such a scenario security and privacy are two key challenges [4] that will determine the success or failure of a connected world.

6.1.1 Motivation

There are various applications of IoT like shopping, education, travel, healthcare, entertainment, and transportation. These work cases can be classified into centralized, and distributed work cases. Distributed work cases are used in the applications where people are mobile such as tourists, and drivers. These nomadic users may utilize any of their devices to conduct their task. The nomadic users perform the task remotely for personal, or professional need through the personal device. This chapter illustrates the remote printing scenario to elucidate the theme of distributed work case. Jack is technophile and by profession a salesman. His job requires business travels across the globe. He can access information, and services both private, and professional through his latest "things" developed for IoT. One of Jack's business trips, Jack uses his handheld device (e.g. Mobile or PDA) to print picture on his home printer while he is away from home. This service is referred as Remote Printing Service (RPS) in this chapter. Jack takes a surrounding scene picture from his personal device when he is travelling around the city, and he wants to share these pictures with his family at the same time. Coincidently, he finds a public- Access Gateway (AGW) in the vicinity which discovers his mobile device via Bluetooth. Jack sends a request requesting for services from the public AGW. As his mobile device is not registered on this public AGW, the gateway will first register this device, and will generate service discovery request to Jack's home AGW at home. The home AGW will discover services available at the home network, and send service list to the public AGW. The public AGW then forwards this service list to Jack's mobile device, and also stores this service list combined with some information related to Jack (E.g. Mobile device MAC address) in its database for future use. Now Jack can select the printing service from the list displayed on his mobile device. After this, he is informed to select the picture file, which he wants to print. Finally the picture will be delivered to the home AGW for printing purpose. This scenario clearly explains that authentication, and access control is very crucial in this distributed work case. Proper access control solution in place will ensure that the correct operation is performed on the correct resource, or service. This distributed work case for RPS is depicted in the form of use case in Figure 6.1, and 6.2. Figure 6.1 shows the use case for gateway registration, and Figure 6.2 shows use case for RPS.

6.1.2 Access Control in Internet of Things

Use of IoT is likely to reach more than 41 million end points before the end of 2014. With the increasing user demands and advancements in communication technologies, IoT networks should be reliable and robust which is why the distributed architecture became essential to IoT, and thus leads to the term of distributed IoT. Moreover, with the help of robotic system, i.e. mobile robots, IoT can be extended to support applications in hazardous environment. However, among other things, IoT applications under various environments are subject to security challenges especially in authentication and access control. Due to resource constraints of sensors, it is necessary for security solutions to be very optimized and efficient and furthermore such solutions

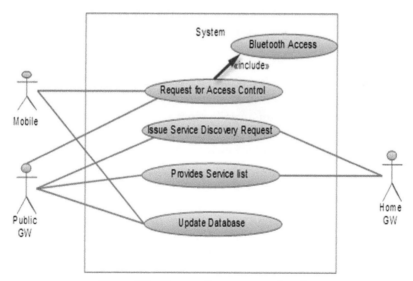

Figure 6.1   Use case for gateway registration

132 Access Control

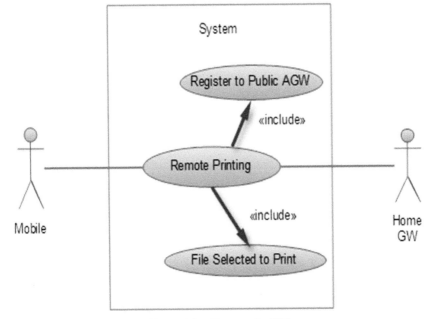

Figure 6.2 Use case for RPS

might differ depending on the application and network architecture. The performance of a security solution is closely related to factors such as generality and granularity. When ubiquitous entities are accessing the IoT, access control is the prime concern as, in IoT applications, access to these distributed and remote nodes and its control is daunting. The task of secure access to IoT resources/devices is a challenging issue. Major factors of influence are the connectivity, power sources, form factor, security, geographical factors and cost of deployment and operation. Applications with different constraints on these factors will have different optimum architectures for integration. IoT design aims at addressing research topics such as designing fundamentally secure access control and authentication, location and IdM. We believe that security cannot be included as an add-on but that there is a need for a robust and coherent architecture ensuring access control and authentication of not only the users but also the ubiquitous devices in IoT. It is also important to understand the relation between security and privacy. Built-in security assuming all users and devices to be trustworthy is essential for IoT applications. Existing protocols are not designed to take care of authentication and access control.

There is concurrent communication for more than one device. IoT is generally a creation of many devices working in part towards common objective and in part for private goals. Successful IoT computing must include undesired interference among communications. Many communication shares pool of resources/devices in a flexible way. There is a need of efficient use of resources. Individual communication varies widely in their demands for computing resources in the course of time. IoT must have mechanism through which communication should request and release resources/devices as per need.

6.1.3 Different Access Control Schemes

Traditionally, access control is represented by an Access Control Matrix (ACM), in which the column of ACM is basically a list of objects, or resources to be accessed and the row is a list of subject or whoever wants to access the resource. From this ACM, two traditional access control models exist, i.e. Access Control List (ACL) and capability-based access control. Many literatures, e.g. [5, 6] have done some comparisons between ACL, and capability-based access control, and the conclusion is that, ACL suffers from a confused deputy problem, and other security threats while this is not the case in the capability-based access control. Moreover, ACL is not scalable being centralized in nature, and also it is prone to single point of failure. It cannot support different levels of granularity, and revocation is time consuming with lack of security. Nowadays, when the Internet, and web-based applications are widely used, different types of access control models have appeared, such as the Role-Based Access Control (RBAC), Context-Aware Access Control (CWAC), policy-based access control, etc. Among others, RBAC is considered to be the most famous access control method in terms of the usage, and implementation in various systems. On the other hand, as mentioned in [6], the RBAC model is essentially a variation of identity-based access control to whom ACL is sometimes referred, which seeks to address the burdens of client identification. Therefore, the RBAC model is still vulnerable to confused deputy problem as the case of the ACL-based model.

In CWAC [7], surrounding context of subject, and/or object is considered to provide access. Scalability is again the problem with CWAC. Delegation, and revocation is not supported at the fullest in CWAC. In CRBAC [8], context is integrated with RBAC dynamically. Context is defined as characterization of surrounding entities for performing appropriate actions. Improper association of context, and role results into scalability, and time inefficiency. Further the delegation is not simple due to context dependency. There are many examples like context-aware patient information system, and context-aware music player where applying role-based access control is a cumbersome process. The comparisons of these access control models are shown in following Table 6.1. They are based on functional parameters such as generic nature, scalability, granularity, delegation, time efficiency, and security.

6.1.4 State of the Art

Many literatures [5, 6, 9] have done detail analysis, and comparisons between traditional access control, and Capability-based Access Control (CAC) and the conclusion is that ACL suffers from a confused deputy problem, and other security threats while that is not the case in the CAC. Moreover, ACL is not scalable being centralized in nature, and also it is prone to single point of failure. It cannot support different level of granularity, and revocation is time consuming with lack of security. However, several drawbacks have been identified in applying the original concept of CAC as it is. [10] Pointed out two major drawbacks of classical CAC namely the capability propagation, and revocation, and provided solutions to them by proposing identity-based capability. Yet, [10] did not clearly describe the security policy that is used in the capability creation, and importantly it did not consider IoT for access control.

There are several access control models of IoT that are inspiring to initiate and extend the research. Recent NIST [11] gives detailed assessment of all access control approaches but besides these established approaches, there are several applications, and scenario specific access control schemes that have been developed. Extended role-based access control model for IoT by incorporating the context information is presented in [12]. In [12], the authors have considered IoT users rather than devices. Furthermore, presented model has been demonstrated with the case studies than implementation. A decision algorithm which is an extension to attribute-based access control with trajectory-based visibility policies is presented in [13]. This is a centralized

Table 6.1 Comparison of different access control models

Models	Generic	Scalable	Granular	Delegation	Time Efficient	Security
ACL	YES	NO	NO	NO	NO	NO
RBAC	NO	NO	YES	YES	NO	NO
CWAC	YES	NO	YES	NO	NO	NO
CRBAC	YES	NO	YES	YES	NO	NO

access control solution for mobile physical objects precisely addressing data access for supply chain management applications. But the secure communication over the network is assumed in [13] which are not practically possible in dynamic scenarios of IoT. High level research on access control, and security management is presented in [14], but the implementation details, and feasibility issues are not discussed. Location-based access control for data security in mobile storage device is presented in [15]. This solution only addresses indoor scenarios, and solutions are again centralized in nature, and not suited for dynamic, and distributed application of IoT. The access control policies based on the usage control, and fuzzy theory is presented in [16], but the practical solution as well as feasibility is left unaddressed. Rule-based context-aware policy language for access control of data, and its prototypical implementation is presented in [17]. This solution is applicable for Electronic Product Code (EPC) information service, and device-to-device access control is not considered. In [8], Context-aware Role-Based Access Control (CRBAC) scheme is presented where context is integrated with role-based access control dynamically. There are many examples like context-aware patient information system, and context-aware music player where applying role-based access control is a cumbersome process. In addition to this, RBAC scheme presented in [18] is not flexible, and does not scale well. As flexibility, and scalability are two important aspects of IoT, this scheme is inappropriate for IoT scenarios. Attribute-Based Access Control (ABAC) schemes presented in [19, 20] are having security issues like confuse deputy problem, and access control management is complex.

Related works shows that existing access control models do not address issues like scalability, time efficiency, and security which are of prime importance in order to apply it to IoT. For any access control scheme in place for IoT, security is the most important issue due to unbound number of devices, and services. This chapter proposes novel, and secure approach of access control for IoT resources i.e. CAC with scalability. Most important design issues of IoT are the scalability, and mobility of heterogeneous devices and CAC will work efficiently for this need.

6.2 Capability-based Access Control

The capability [21–23] is as a token, ticket, or a key that gives the possessor permission to access an entity, or object in a computer system. Conceptually, a capability is a token that gives permission to access an object. In the context of IoT, an object is a device, service, or any object quipped with RFID tags.

136 *Access Control*

A capability is implemented as a data structure that contains two items of information: a unique object identifier, and access rights. The access rights define the operations that can be performed on that object. Examples of capability are: a movie ticket is a capability to watch the movie, and a key is a capability to enter house. Using capabilities we can name those objects for which a capability is held, and it also achieves the least privilege principle [24]. Capabilities have been implemented as lightweight access control in many OS and distributed environments [25, 26].

6.2.1 Concept of Capability

There is large research done in the area of access control. Traditionally, access control is represented by an Access Control Matrix (ACM), in which the column of ACM is basically a list of OBJECTS, or resources to be accessed, and the row is a list of SUBJECTS, or whoever wants to access the resource. From this ACM, two traditional access control models exist, i.e. Access Control List (ACL), and Capability-based Access Control (CAC). Due to unbound number of devices, and services, scalability, and manageability issues are daunting in IoT. With the increasing complexity, ACL, or Capability List (CL) are widely used for access control solutions. ACL presents column view, and CL presents row view of access control matrix. CL is attached to the device, and specifies its related services, or resources. Each entry in CL is capability which is a pair of service/resource, and set of access rights. Conceptually, a capability is a token, ticket, or a key that gives permission to access a device. Figure 6.3 sketches the main difference between ACL, and CAC models [27].

Figure 6.3 explains that CL can prevent confuse deputy problem [28], and achieves principle of least privilege. According to Figure 6.3, the arrows for ACLs direct from the resources/services to devices but the arrows for CL direct from devices to the resources/services. This means that the capability pairing between devices and resources/services is generated by the system. Thus, the permission of devices to access resources/services can be modified by the built-in methods. Oppositely, the system with ACL approach must need a special method for pairing devices to resources/services. This is the first advantage of capability over the ACL.

A capability is a token that gives permission to access device. A capability is implemented as a data structure that contains two items of information: a unique device identifier, and access rights. For simplicity, it is sufficient to examine the case where a capability describes a set of access rights for the device. The device may also contain security attributes such as access rights, or other access control information.

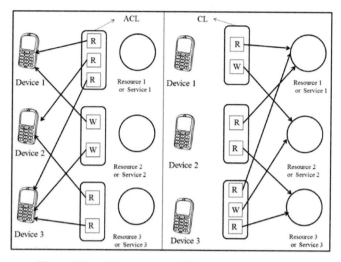

Figure 6.3 ACL versus capability-based access control

6.2.2 Identity-based Capability Structure

However, several drawbacks have been identified in applying the original concept of CAC as it is. [10] Pointed out two major drawbacks of classical CAC namely the capability propagation, and revocation, and provide solutions to them by proposing identity-based capability. Yet, [10] did not clearly describe the security policy that is used in the capability creation, and importantly it did not consider IoT for access control.

For simplicity, the capability describes a set of access rights for the device. The device which may also contain security attributes such as access rights or other access control information. Identity-based Capability (ICAP) structure is shown in Figure 6.4 with how capability is used for access control. ICAP is represented as shown in Equation (6.1)

$$ICAP = (ID, AR, Rnd) \qquad (6.1)$$

Where

- *ID*: Device identifier
- *AR*: Set of access rights for the device with device identifier as *ID*
- *Rnd*: Random number to prevent forgery and is a result of one way hash function as given in Equation (6.2)

$$Rnd = f(ID, AR) \qquad (6.2)$$

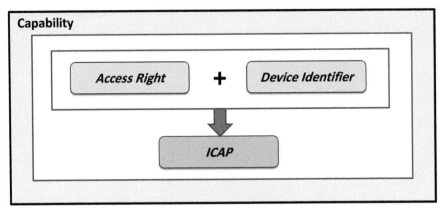

Figure 6.4 Identity driven capability structure

Where f is a publicly known algorithm based on public key cryptosystem to avoid the problem of key distribution. When the device receives access request along with the capability, one way hash function is run to check the *Rnd* against tampering. If the integrity of the capability is maintained, then access right is granted. Capability structure is depicted in Figure 6.4. This capability is not stored centrally on a particular device. Each device has its own capability which is verified by each access. First, both the devices get connected to ad-hoc network and then an identity is generated for these devices based on media access control address for unique identification. After this, the connection requests are sent, and the connection is established. The access rights are decided, and capabilities are created for these devices. The capabilities are exchanged along with a message digest. SHA-1 message digest is used to check the tampering, or forgery of capabilities.

6.2.3 Identity-driven Capability-based Access Control

Identity-driven capability-based access control works in two stages. In first stage, the devices are connected to each other and in the second stage, capability-based access is allowed to other device through ICAP. Each communication that is to be established is verified by its capability access. Only after the capability verification, the devices are able to communicate with each other. Any device that wants to communicate with the other device is able to initiate the communication by sending the request to a specific device. The next stage is to verify whether that requesting device is having the capability to communicate with the called device. This access right gets checked using

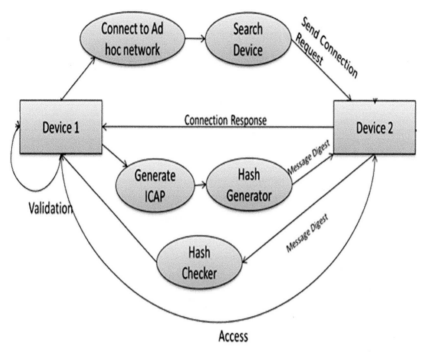

Figure 6.5 High level functioning of CAC

the capability of that device which is associated with every device. To send the capability, message digest using SHA-1 is generated for each device as stated earlier, and the remote device will check its validity using SHA-1. Figure 6.5 shown depicts high level functioning of CAC.

6.3 Implementation Considerations

Complete CAC scheme is presented in Figure 6.6. Figure 6.6 shows access based on CAC between two devices. All devices are treated as subjects and resources to be accessed as objects. In this implementation of CAC, file is considered as object for access. Access rights (AR) is shown below in Equation (6.3).

$$AR \in \{Read, Write, NULL\} \qquad (6.3)$$

AR can either be {Read}, {Write}, {Read, Write}, or {NULL}. If AR = {NULL}, the permission to access particular object is not allowed.

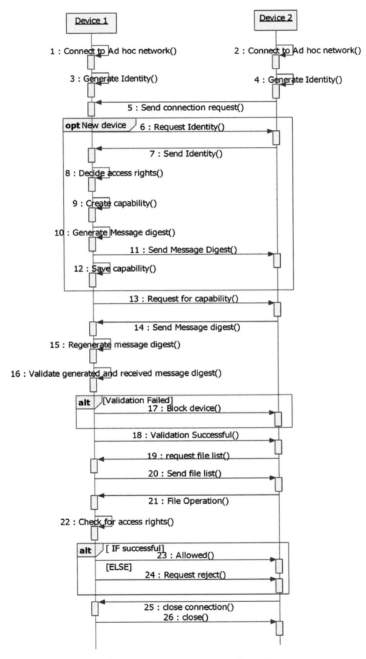

Figure 6.6 CAC scheme for IoT

Once the capability is verified against forgery, both the devices are able to perform operations as specified in capability, and access is granted. As any device can perform only those operations as specified in capability, the principle of least privilege is supported to a large extent.

CAC can be implemented in five modules which are described below:

Data Exchange: As the name suggests, the main purpose of data transfer module is transfer of data between two connected devices. Data exchange is done according to the access rights specified in capability.

Hash Handler: Hash handler works with the one way hash function using SHA-1. We can use one way hash function to store the capability in remote device. The generated message digest will be transferred to the device, and for each data communication the same digest is used for communication. This is useful for ensuring the modification in the identity capability.

File Browser: File browser module helps to show the directory structure of the remote device to which the connection is established, and the data transfer is to be done. When any connection is made to the remote device, file browser fetches the files from the directory of remote device. File browser is nothing but the list showing the directories of remote device using a connected device which can access the required files according to its access rights

Initializer: Initializer initializes the application, and it checks for the ad-hoc network connectivity.

Device Discovery: Device discovery module discovers the devices which are in the range for communication. Device discovery shows the list of the devices after searching to which it can connect for communication.

6.3.1 Functional Specifications

Different use cases in the CAC scheme are shown below in the Figure 6.7, 6.8, 6.9, and 6.10. Use case for connection establishment between two devices is shown in the Figure 6.7 in which the system includes all the steps in connection establishing process.

Use case for ICAP generation is depicted in Figure 6.8. Deciding the access rights, ID generation, and generation of the capability are the major tasks in generating ICAP as shown below. System includes these three steps for ICAP generation.

Figure 6.9 shows use case for sending ICAP from one device to another device. The process of sending ICAP includes getting ICAP, and generating hash for that ICAP, and the complete system is shown below.

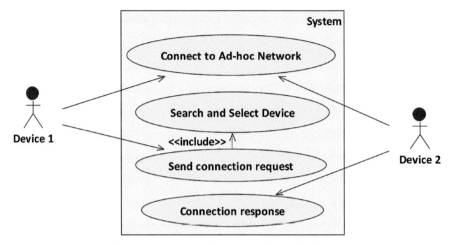

Figure 6.7 Use case for connection establishment

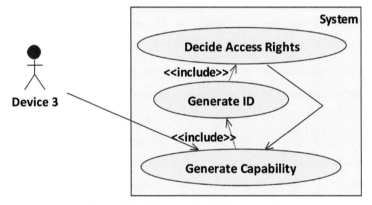

Figure 6.8 Use case for ICAP generation

Receiving ICAP is the main step in CAC, and the corresponding use case is shown in the Figure 6.10. Receiving ICAP includes checking of hash, and access validation, and once the validation is done only then the access is granted.

6.3.2 Access Control Policies Modeling

Controlling access to information or resources is usually done by defining access control rules, which decide who is allowed to access what and who

6.3 Implementation Considerations 143

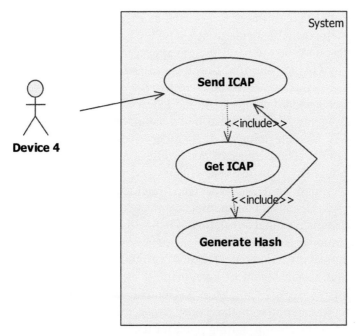

Figure 6.9 Use case for sending ICAP

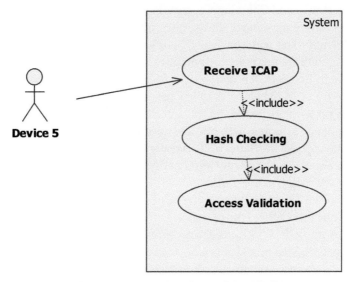

Figure 6.10 Use case for receiving ICAP

is not. These rules take different forms such as RBACs, ACLs, policies etc. Before the development of standards based policy languages, interoperability was a major concern. It was with the emergence of the XACML proposal [29], defined by OASIS, that IdM developers started thinking about how to make use of such standards based languages to define the set of policies, and to provide more standard solutions. In the IoT world, such standards based solutions are imperative due the distributed nature of the problem. XACML includes an XACML delegation profile in order to support administrative and dynamic delegation. The purpose of this profile is to specify how to express permissions about the right to issue policies and to verify issued policies against these permissions. This profile, lead to an identity federation scenario, which is the key element upon the management of delegation policies. At the moment there is no suitable solution to define the relationship among the involved institution in a service interaction, nor a way to combine the decision taken by different organizations. There is currently no standard proposal related with the establishment of agreement at organization, federation or other trust domains levels. Examples of this kind of policies could be common information representation format, security requirements, levels of trusts, etc. This policy can be taken as a starting point for the definition of a negotiation mechanism about capabilities and policies, independently of the kind of entity involved on it. Policy and Charging Control (PCC) in LTE enables centralized mechanism for charging control and service-aware quality of service. PCC operates in S9 interface and consist of Policy and Charging Rule Function (PCRF) which controls the policies dynamically based on subscriptions and sessions between home PCRF and visited PCRF. Consider the scenarios of heterogeneous home M2M network in IoT based on LTE /4G. In this scenario, home gateway proactively and adaptively interacts with the surrounding radios in order to connect to home network and in turn to the external networks. Security policies protect the home M2M network from possible external attack via trusted access control and networked encryption technique.

Although XACML was the starting point towards the definition of standard policies, it is only focused on the resource access control type of policy. More or less at the same time, other kind of policies emerged to cover specific aspects for IdM, for example P3P [30], to define online privacy release information policies between end users and services. Current systems have incorporated these kinds of standard policies in some way, for example Shibboleth [31] and Liberty Alliance [32] providing definition of access control policies by means of XACML. But there is a need to define policies in a standard way in the next generation of policy-driven systems when distributed scenarios in the IoT domain are considered.

6.3.3 Mobility and Backward Compatibility

Furthermore, there are few challenges to implement CAC in mobile environment. Access delegation method with security considerations based on capability based context aware access control scheme intended for federated IoT networks is presented in [33]. In [33], capability propagation incorporating context in federated IoT environment with scalability and flexibility for distributed systems is presented. Authority delegation for mobile and federated environments is challenging due to its dynamic and distributed nature. Another issue is that, it is necessary to have an established trust relationship between all entities prior to delegation. As the CAC is addressing device-to-device authentication and access control, it is compatible in the user equipment and network elements being light-weighted and flexible in nature. Backward compatibility with the legacy network should not be the issue with the availability of high and powerful resources. In a mobile environment, mobility management is an interesting issue to deal with. The A interface which is an interface between mobile switching service switching centre and base station system which support many application part and Direct Transfer Application Part (DTAP) is one of them. Mobility management is one of the functionality of DTAP. There are many mobility management messages which are exchanged for identity establishment and access control (AUTHENT_REJ, AUTHENT_REQ). As physical layer of the A interface is 2 Mbps digital connection and DTAP deals with the exchange of layer 3 messages, no major adaptations are required to make CAC functional.

As presented in [34], wireless communication and evolution is being faced by many constraints. These constraints are regulatory constraints like operating rules on the communication device and pre-decision on the frequency bands. Layered design of the communication protocol introduces architectural constraints which is important for proliferation. Other constraints are standardization constraints in which the particular communication protocol is developed and operated. The backward compatibility also needs many refinements and technological improvements for new standards. There are also market and social constraints deals with the new applications and the requirements from communication systems. Figure 6.11 depicts the outline of the evolution in wireless communications. As shown in the Figure 6.11, ws1 and ws2 get converged and system ws5 is emerged. When ws4 is

146 *Access Control*

evolved, it is not feasible to implement concept c2 due to heavy constraints as discussed above, but due to increasing requirements (by ws3 also) the constraints are refined to change and ws7 is evolved. Over the period of time, some of the wireless communication systems become obsolete. Example of this obsolete system is shown in the Figure 6.11 which happens for ws2. The important thing to be noted here is that the constraints do not allow the concept c3 to be implemented over the period of time frame as depicted in the Figure 6.11.

Similar to global Internet scenario, interoperability and internetworking is ensured by following OSI stack but still there are many exceptions due to unpredictable nature of wireless interface. This makes it more difficult to guarantee expected quality of service in resource constrained IoT and next generation networks. Backward compatibility to legacy networks is a challenge due to lack of cross layer coordination which is a need of today in order to get performance improvement. Other interoperability and

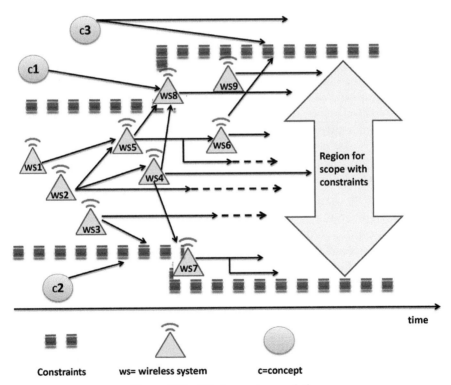

Figure 6.11 Wireless system evolution

internetworking issues are architecture design and multi-traffic environment. To address these ensuing issues, there is a need for lot of research.

6.4 Conclusions

There is concurrent communication of more than one device in IoT in private or public domain. Successful IoT communication and computing includes sharing of pool of resources / devices in a flexible way. For this purpose, there is a need of secure access of resources. Access control and authorization in IoT with the least privilege is very important to establish secure communication between multiple devices, and services. Access control is a critical functionality in IoT, and it is particularly promising to make access control secure, efficient and generic in a distributed environment. Another important property of access control system in the IoT is flexibility which can be achieved by access control. Access control is of paramount importance and full thrive of IoT, especially due to dynamic network topology and distributed nature. Centralized and distributed IoT work cases are presented and discussed initially in this chapter to understand the need of secure access control. This chapter also presents study of different access control models with their advantages and limitations.

The capability-based authorization approach for management of access control to set of devices, and services is introduced in this chapter. Access control issues, and challenges specific to IoT are presented and explained. The concept of capability and its application for access control in IoT is the main contribution of this chapter. As stated earlier, a simple example of capability used in the real life is a ticket that allows a passenger to embark on different modes of city transportation, e.g. bus, tram, train, etc. If the ticket contains all the transportation modes in which the passenger can embark, then the capability-based access control is used. If the capability that is presented by the Subject matches with the capability that is stored in the device or an entity that manages the device, access is granted. By using the identity and capability-based access control approach together, this proposed model provides scalability, flexibility, and secure authority delegation for highly distributed system. However, unlike the classical capability based system, ICAP introduced the identity of Subject or User in its operation. In this way, it claimed to reduce the number of capabilities stored. Key point to be noted is that the CAC achieves principle of least privilege by selective disclosure.

In next part of this chapter, implementation considerations for CAC have been discussed with different possible modules as well as use cases. Functional specifications are also presented to get a more clear idea of CAC. Controlling access to information or resources is usually done by defining access control rules, which decide who is allowed to access what and who is not. Access control policies modeling based on XACML is discussed. In the IoT world, such standards based solutions are imperative due to the distributed nature of the problem. Equally it is important to understand that the proposed CAC is compatible with the state of the art, both from the perspective of the UE and the network (Access and Core) elements. Also is it backward compatible with legacy networks? In the sequel, next part of this chapter discusses backward compatibility with state of the art as well as legacy networks. At the end, challenges to implement the new CAC in a mobile environment are presented and discussed.

However, in IoT, heterogeneity is an important property; therefore it is difficult to comment that the generic solutions exist as the proposed work did not consider RFID networks. Most of the mechanisms proposed in this contribution are based on the assumption of synchronization. Authentication, access control, all these operations are based on the time-stamps exchanged between nodes. However, this is a rather strong assumption, and we have not addressed the problem behind synchronization. Future work will also focus on achieving authentication combined with access control and its analysis for adversaries as well as prototype implementation with the existing access control realization model, such as XACML. Future work will include assigning risk to different threats to obtain quantitative analysis and finding techniques for threat mitigation. Future research should seek device level implementation for other security dimensions which may provide low cost solutions. Finding realistic solutions that balance security and efficiency is a significant research area.

References

[1] K. Ducatel, M. Bogdanowicz, F. Scapolo, J. Leijten, and J-C. Burgelman, "Scenarios for Ambient Intelligence in 2010," IST Advisory Group (ISTAG), European Commission, (Brussels, 2001).

[2] Association for Computing Machinery (ACM), "The Next 1000 Years," Special Issue of Communications of the ACM, 44: 3 (2001).

[3] MAGNET Consortium, MAGNET: My Personal Adaptive Global NET, Integrated Framework Programme, Information Society Technologies (IST), Nov 2003.

[4] Mayer, C., "Security and Privacy Challenges in the IoT", WowKivs, Electronic Communications of the EASST, Volume 17, 2009, Germany.
[5] Ravi S. Sandhu, "The Typed Access Matrix Model," In Proceedings of the IEEE Symposium on Security and Privacy 1992, IEEE CS Press, USA, pp: 122–136.
[6] T. Close, "ACLs don't," HP Laboratories Technical Report, February 2009.
[7] Y. G. Kim, C. J. Mon, D. Jeong, J. O. Lee, C. Y. Song, and D. K. Baik, "Context-Aware Access Control Mechanism for Ubiquitous Applications," In the Proceedings of Third International Conference on Advances in Web Intelligence, LNCS, Volume: 3528/2005, pp: 236–242, Lodz – Poland, June 6–9 2005.
[8] D. Kulkarni, and A. Tripathi, "Context-Aware Role-based Access Control in Pervasive Computing Systems," SACMAT'08, pp: 113–122, Estes Park, Colorado, USA June 11–13, 2008.
[9] M.Miller, Ka-Ping Yee, and J. Shapiro, "Capability Myths Demolished," Technical Report SRL2003–02, System Research Laboratory, Johns Hopkins University, 2003.
[10] Gong, L., "A Secure Identity-based Capability System," In Proceedings of IEEE Symposium on Security and Privacy, pp: 56–63, IEEE Computer Society Press, Los Alamitos, Oakland –CA, May 1–3 1989.
[11] Vincent C., Hu, D. F. Ferraiolo and D. Rick Kuhn, "Assessment of Access Control Systems," Interagency Report 7316, National Institute of Standards and Technology, Gaithersburg, MD 20899–8930, September 2006.
[12] Guoping Zhang, and Jiazheng Tian, "An Extended Role based Access Control Model for the Internet of Things," In International Conference on Information Networking and Automation (ICINA), 2010., Volume: 1, no., pp: V1-319–V1-323, Kunming – China, October 18–19 2010.
[13] Florian Kerschbaum, "An Access Control Model for Mobile Physical Objects," In Proceedings of the 15th ACM symposium on Access control models and technologies (SACMAT '10). ACM, New York, USA, pp: 193–202, Pittsburgh, PA, USA, June 9–11 2010.
[14] Kun Wang, Jianming Bao, Meng Wu and Weifeng Lu, "Research on Security Management for Internet of Things," In International Conference on Computer Application and System Modeling (ICCASM), 2010. Volume: 15, no., pp: 133–137, Taiyuan, October 22–24 2010.

[15] Zhang Xin-fang, Fang Ming-wei and Wu Jun-jun, "An Indoor Location-based Access Control System by RFID," In IEEE International Conference on Cyber Technology in Automation, Control, and Intelligent Systems (CYBER), 2011., pp: 43–47, Kunming-China, March 20–23 2011.

[16] Guoping Zhang and Wentao Gong, "The Research of Access Control Based on UCON in the Internet of Things," Journal of Software, Volume: 6, No 4 (2011), 724–731, April 2011 (JSW, ISSN 1796-217X) Copyright @ 2006–2012, Academy Publisher.

[17] E. Grummt, and M. Müller, "Fine-Grained Access Control for EPC Information Services," In Proceedings of the 1st International Conference on The Internet of Things- IOT 08, 2008., Volume 4952 of LNCS, pp: 35–49, Springer-Verlag Berlin Heidelberg. Zurich.

[18] INCITS CS1.1 RBAC Task Group, "INCITS 459 Information technology– Requirements for the Implementation and Interoperability of Role Based Access Control (RBAC)," Draft, August 2010.

[19] E. Yuan and J. Tong, "Attributed Based Access Control (ABAC) for Web Services," In Proceedings of the IEEE International Conference on Web Services (ICWS'05). Florida – USA, July 11–15 2005.

[20] D. R. Kuhn, E. J. Coyne and T. R. Weil, "Adding Attributes to Role–Based Access Control," In IEEE Computer Journal, Volume 43, Issue: 6, pp: 79–81 June 2010.

[21] Parikshit N. Mahalle, Bayu Anggorojati, Neeli R. Prasad, and Ramjee Prasad, "Identity Establishment and Capability-based Access Control (IECAC) Scheme for Internet of Things," In proceedings of IEEE 15th International Symposium on Wireless Personal Multimedia Communications (WPMC – 2012), pp: 184–188, Taipei - Taiwan, September 24–27 2012.

[22] Parikshit N. Mahalle, Bayu Anggorojati, Neeli R. Prasad, and Ramjee Prasad, "Identity driven Capability-based Access Control (ICAC) for the Internet of Things," In proceedings of 6th IEEE International Conference on Advanced Networks and Telecommunications Systems (IEEE ANTS 2012), Bangalore – India, December 16–19 2012.

[23] J. B. Dennis and E. C. van Horn, "Programming Semantics for Multi-programmed Computations," In Communications of the Association for Computing Machinery, Volume: 9, Issue: 3, pp: 143–155, March 1966.

[24] J. H. Saltzer, and M. D. Schroeder, "The Protection of Information in Computer Systems," In Proceedings of the IEEE, Volume: 63, Issue: 9, pp: 1278–1308, September 1975.

[25] H. M. Levy, "Capability-Based Computer Systems," Digital Press, Bedford, MA, USA, 1984. http://www.cs.washington.edu/homes/levy/capabook/.
[26] J. S. Shapiro, J. M. Smith, and D. J. Farber, "EROS: a Fast Capability System," In ACM Operating Systems Review, Volume: 33, Issue: 5, pp: 170–185, December 1999, Proceedings of the 17th Symposium on Operating Systems Principles (17th SOSP'99).
[27] M. Stamp., "Information Security Principles and Practice," John Wiley & Sons Inc., NJ. 2006.
[28] N. Hardy, "The Confused Deputy: (or why capabilities might have been invented)," In ACM SIGOPS Operating Systems Review, Volume: 22 Issue: 4, pp: 36–38, October 1988.
[29] OASIS. eXtensible Access Control Markup Language (XACML) Version 3.0, February 2009. Working Draft 8.
[30] W3C Platform for Privacy Project: http://www.w3.org/Privacy/.
[31] The Shibboleth project –www. shibboleth.net.
[32] The Liberty Alliance Project - www.projectliberty.org
[33] Bayu Anggorojati, Parikshit N. Mahalle, Neeli R. Prasad, and Ramjee Prasad, "Capability-based access control delegation model on the federated IoT network". In IEEE 15th International Symposium on Wireless Personal Multimedia Communications (WPMC – 2012), pp: 604–608. Taipei - Taiwan, September 24–27 2012.
[34] Petar Popovski, "On Designing Future Communication Systems: Some Clean-Slate Perspectives". In springer book chapter titled: Globalization of Mobile and Wireless Communications (Editors: Ramjee Prasad, Sudhir Dixit, Richard Nee, Tero Ojanpera), pp: 129–143, 2011, Springer Science+Business Media B.V.

7

Conclusions

7.1 Summary

IoT is a great technological revolution which represents the convergence of Information, communication and technology. The envisaged IoT applications present the vision of ubiquitous computing by making the things smart. The six chapters described in this book and the challenges, trends identified in the respective chapters are crucial for IdM in IoT. The success in the development of wireless communication has resulted in increasing number of technological as well as security challenges for futuristic IoT. Due to exploded use of smart devices, WSN and RFID networks, market competition is increased resulting into lower cost. Secure IdM of things, network entities and smart devices are the core elements of IoT security. IdM comprises set of solutions for addressing, trust management, authentication and access control. This book defined the problem of IdM in IoT which includes identifying things, assigning identifiers to them, performing authentication and managing access control. In IoT, each real thing becomes virtual means that each entity has locatable, addressable and readable foil on the Internet. We need to identify resources, devices, agents, relationships, mappings, properties, and namespaces and provide identity securely. Traditional security solutions will almost certainly not suitable due to resource constraints and scale to IoT's amalgam of context and devices. WSN and RFID are identified as the likely candidates for IoT and scalability, resource constraints and distributed nature of IoT as key challenges to address IdM problem.

It should be noted that most research has focused on IdM issues in the Internet and web computing era by only orienting users. Current IdM solutions are designed with the expectation that significant resources would be available and applicability of these solutions to IoT is unclear. Even the fundamental question of how well the IdM problem in the resource constrained IoT would be solved conceptually? has been given little attention.

By formalizing the IdM framework and the proposals for every building block of the proposed framework can make the device-to-device communication secure. The roadmap and approaches presented in this book has identified some of the important challenges for IdM in IoT. The main challenges with respect to the design issues of resource constrained IoT, and application areas are also discussed. For the identified challenges of IdM, the existing methods, and schemes are investigated, and the new approaches are suggested that can give better results and performance or can give a different outlook to extend these methods.

In the first chapter, this book has considered different application scenarios of IoT in order to understand IdM requirements and challenges. IoT vision, emerging trends and economic significance is discussed to understand the IoT potential. In the sequel, technical building blocks of the IoT are described with the proposed layered architecture. The role of RFID in the IoT from a visionary technological perspective is introduced and illustrated its expected business relevance. We also describe design issues, technological and security challenges to understand how solutions to multiple problem areas in the IoT are to be designed. Finally this chapter concludes with the different IoT application areas like manufacturing, logistic and relays, energy and utilities, intelligent transport, environmental monitoring and home management. This chapter has also introduced example usage scenarios to understand how different stakeholders across different business verticals will benefit from IoT.

Second chapter presents elements of IoT security essentially in the context of IdM. High level view of IoT is presented, and different threats are discussed. This chapter has presented vulnerabilities of IoT along with IoT security requirements. The detailed security architecture with possible threats and attacks are presented in this chapter which provides a systematic way of countering possible threats. This architecture defines, and proposes the framework for IoT applications, and dimensions needed to achieve security for IoT. This attack, and threat analysis gives motivation to IdM in IoT. Threat modelling based on the use cases is provided to understand how exactly attack happens? Identity establishment, access control, data and message security, nonrepudiation and availability are identified a key elements of IoT security.

The era of Internet computing and IoT is in need of reliable IdM mechanism. Chapter 3 introduces the reader to the elements of IdM. IdM and identity portrayal in the IoT context as well as state of the art technologies in the Internet computing and IoT is presented and discussed. This chapter has provided a classification of existing approaches based on how the scope of identity is defined? We have also discussed the benefits and limitations of

7.1 Summary

existing IdM models. This chapter has also presented the elements of federated identity and discussed identity provisioning. The evolution of computing from centralized to distributed era has considerably increased the complexity of IdM. Varied scope of identity results into different IdM architecture like centralized, distributed and hybrid. In the sequel, there are different IdM approaches like user-centric, device-centric and hybrid. Finally, parametric comparison of six different IdM technologies is presented at the end of this chapter.

Connecting ubiquitous devices over public networks requires secure process for establishing a secure context before these devices are attached for particular activity or operation. Identity and trust have been indicated as highly important aspect in the context of IoT networks. Chapter 4 introduces importance of identity and trust in the IoT. Motivation for trust management in the IoT, trust management life cycle and the survey of existing models in identity assurance is presented and discussed in this chapter. IdM approach alone is not sufficient if it is not integrated with trust essentially across federated IoT networks. Various public-key trust models, trust establishment using third party approach and attribute certificate-based trust management is explained in the next part of this chapter. Emerging mechanisms for the exchange of security constructs in the context of web or IoT like: web of trust model, SAML and fuzzy approach for trust management is presented and discussed. At the end, this chapter also explains the importance of trust for IdM and access control in IoT. A relationship between trust and access control is introduced in the last part of this chapter.

Secure entity identification known as authentication is referred as identity establishment in this book. As IoT becomes discretionary part of everyday life, could befall a threat if security is not considered before deployment. Identity establishment is concerned with the scheme or method of secure association between entities or group of entities. In the IoT context, entities represent devices/services/users. The goal of this chapter is to discuss motivation and challenges of identity establishment in the IoT. IoT use case along with the attack modelling for attacks like man-in-the-middle, replay and denial of service is presented to understand how these attacks occur in the IoT context. The detailed survey of different authentication scheme as well as evaluation of these scheme based on few performance parameters is discussed to understand the gap analysis in next part of this chapter. There has been lot of debate about which of the cryptographic primitives like public key or private key is suitable for the IoT. The comparison of public and private key cryptographic primitives and their application to the IoT security is also discussed in this chapter. In the sequel, ECC-based mutual authentication scheme is presented and discussed along with the highlight on different security protocol verification tools at the end of this chapter.

Access control is important and critical functionalities in the context of IoT to enable secure communication between devices. Mobility, dynamic network topology and weak physical security of low power devices in IoT networks are possible sources for security vulnerabilities. Access control issues, and challenges specific to IoT are explained in this chapter. The concept of capability and its application for access control in IoT are the main contributions of this chapter. This chapter has introduced the capability-based authorization approach for management of access control to set of devices, and services. The concept of capability and its application for access control in IoT is the main contribution of this chapter. As a result, this chapter has presented and discussed novel identity-driven capability-based access control for IoT along with the implementation details, functional specifications and access control policy modelling. Issues and challenges related to application of capability concept in the mobile environment and backward compatibility to the legacy networks is also discussed at the end of this chapter.

7.2 Identity Management Framework

This book proposes a secure cross layer collaborative IdM setup to cater to the requirements coming from IoT. The architecture ties the IdM of the service layer together with the security, and access control needed for interactions between the things. The IdM architecture is shown in the Figure 7.1.Things are devices with network capabilities ranging from high-end devices such as mainframes to simple sensors. Firstly, these things will belong to many different user spaces, and they need to be able to collaborate together despite their heterogeneity. In order to achieve this, the book proposes a framework with secure interaction methods for identity establishment observing different access policies in order to fulfil a specific functionality. When talking about functionality of this setup, this book thinks about services, which can be found above this architecture. There are many different services we can think of, i.e. the mainframe may use external temperature sensors to check whether the temperature in the room is above a certain level to trigger an alarm, or more composed services such as the ones gathered under the three scenarios like private, enterprise and e-Health. In the middle of both layers, IdM middleware layer securely manages the relationships between devices/things, and services.

7.2 Identity Management Framework

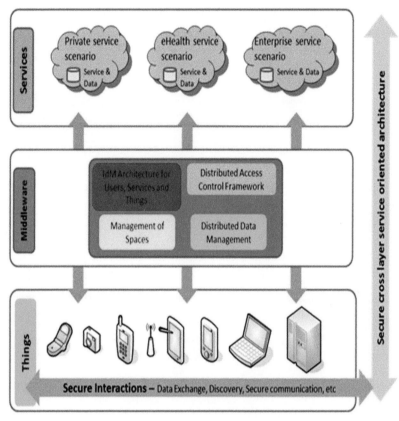

Figure 7.1 IdM architecture [1]

This book also proposes IdM framework extending current IdM architecture and defined new ones for the devices in the network to help users, and devices to interact securely with one another. This framework is depicted in the Figure 7.2. This framework enables devices to communicate with other surrounding devices in environments with different security, and authentication requirements. The authentication feature of the framework covers the authentication of devices, where the relying parties may be services, other things/devices or users. As a final outcome, IdM for IoT devices should be energy efficient, scalable and lightweight solution for every building block of the proposed framework. IdM is achieved based on the trust, context-aware addressing, authentication and access control. Identity/location privacy of the user would have been another building block in the proposed framework to address IdM of the user as well as devices.

158   *Conclusions*

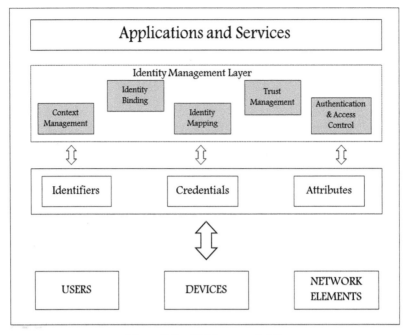

Figure 7.2   IdM framework [1]

7.3 Future Outlook

There are still many aspects of IdM that were not considered in this book due to the fact that some of them were out of scope or due to limitation of time and resources. There are plenty of rooms that can be explored and added on top of our proposed framework. Beyond the issues that have been evaluated in the scope of this book, there are still numerous aspects for further research. Here, we would like to add some of these possibilities and open issues that came across our mind while working over this book. At the end of this book, we point out some thoughts, and open problems for future research:

We believe that IdM itself is a very big administrative domain, and requires a lot of attention in the future to provide more scalable and complete solutions. It seems that all security protocols are limited by their requirement regarding computational efficiency and scalability due to unbound number of devices. It would be valuable to have more formal analysis for these limitations. The formal analysis would include designing formal specifications and semantics in order to build a complete solution. It would be further an interesting approach to address identity/location privacy of the user and integrate it in the proposed IdM framework. Location privacy is equally important risk in IoT. To ensure location privacy, communication and reference signal integrity needs to be maintained. Communication confidentiality and privacy of localization and tracking data is highly sensitive in IoT amalgam. Therefore location privacy is indeed an important issue to address further which include ensuring the privacy of localization data of user as well as devices.

Authority delegation would be another interesting extension of the proposed CAC model to look forward. The future outlook will consider the case in which no prior knowledge of the trust relationship between two network domains in Federated IoT. Future work will involve specification as well as security evaluation of the CAC propagation and revocation in order to have a complete model and verification of CAC mechanisms. Complete interoperability and internetworking is still an open research area to take this research further.

As a next step, it could be evaluated how the proposed extensions of the fuzzy approach for trust-based access control can be applied to multi-contexts scenarios using weighted averaging operator. Beyond IoT, integration of this trust model in Web 2.0 seems to be more promising with the real adversaries like know thy enemy. Evaluation and comparison of the different trust-based access control schemes integrated with Web 2.0 will be another interesting area to explore.

Furthermore, it would be interesting to integrate context, and trust together to get a context-aware trust management, and extend the evaluation of this for the trustworthiness of the group of entities. A research is also needed to evaluate the performance, and security effectiveness of the proposed authentication, and access scheme on RFID that incorporates dynamic context information. It would be interesting to extend IdM framework to incorporate heterogeneity of the devices. Another interesting extension of this research would be test proposed IdM framework in the converge network.

Reference

[1] Parikshit N. Mahalle (2013). Identity management framework for Internet of Things. Ph.D. Thesis. CTIF, Aalborg University: Denmark.

Index

A
Access Control 29, 33, 48, 110
Addressing 7, 63, 104, 157
Attacks on IoT 1, 18, 31, 44, 79
Attribute Certificates 85, 87
Authentication 9, 69, 73, 103, 110
Authentication and Access Control 9, 69, 109, 131, 157
Authorization 17, 72, 73, 94, 156
Availability 33, 47, 49, 109
AVISPA 116, 119, 120, 124

C
Capability 1, 89, 115, 133, 137
Capability-based Access Control 48, 115, 135, 156
Context Management 4, 52
Cryptography 48, 86, 109, 111, 123
Cryptosystem 48, 108, 109, 138

D
Data and Message Security 47, 48, 154
Data confidentiality 33
DDoS 32, 51
Device-centric Identity Management 71
DoS attack 31, 42, 105, 107, 124

E
Elliptic curves 125
EPCglobal Network 13, 16
EPCIS 13, 15, 16

F
Features of the future IoT 9, 25
Federated identity 64, 67, 155
Flexibility and adaptability 34

Fuzzy Approach for Trust 85, 95, 98, 155, 159
Fuzzy Rule Base 82, 95

G
Global identity 64, 69

H
Hardware threat 46
Hierarchical Addressing 10, 26, 63, 75, 76
Hybrid Identity Management 71

I
Identifiers in IoT 56, 57
Identity Establishment 22, 65, 103, 104, 156
Identity Management 79, 132
Identity Management Framework 156, 159
Identity Management in IoT 55, 56, 75
Identity Management Models 55, 64, 79
Identity Portrayal 59, 60, 74, 154
Identity-based Capability Structure 137
IdM architecture 155, 156, 157
Information corruption 32, 34
Information disclosure 32, 34
Internet of Things 1, 2, 29, 69, 150
IoT Layered Architecture 2, 9, 11, 154
IoT Security Tomography 43, 44
IP for Things 17

L
Local identity 64, 65, 85

M

Man-in-the-middle attack 32, 40, 105, 122
Mutual Authentication 48, 109, 113, 124, 156
Mutual Identity Establishment in IoT 104

N

Network identity 64, 66, 67
Non-repudiation and Availability 49, 51

O

Object Classification 52
One Way Authentication 113, 116, 124
ONS 15, 16, 17
OS threat 46

P

PKI trust topologies 87, 88, 97
Privacy 1, 37, 80, 121, 144
Private Key Cryptography 111
Public Key Cryptography 48, 87, 111, 120
Public Key Infrastructure 83, 85, 86
Replay attack 106, 124
RFID 61, 103, 110, 148

S

SAML Approach 94
Scalability 7, 18, 20, 109, 136
Secret Key Generation 113
Secure software execution 33

Secure storage 34, 36
Security 31, 35, 43, 47, 62, 103
Security Analysis 84, 118
Security architecture for IoT 34, 118
Security Protocol Verification Tools 116, 156
Sensor threat 46

T

Tamper resistance 34
Theft of resources 32, 33, 34
Threat Analysis 29, 37, 121, 154
Threats 31, 45, 133, 154
Threats on MAC layer 45
Threats on network layer 45
Threats on RF layer 46
Threats on transport layer 44
Trust 21, 63, 79, 82, 155
Trust management 82, 153, 155, 159
Trust Management Life Cycle 82, 83, 97, 155

U

Unauthorized access 31, 32, 34, 38
User-centric Identity Management 70

V

Vulnerabilities of IoT 29, 31, 154

W

Web of Trust Models 90, 98
Web Services Security 92, 93

About the Authors

Parikshit N. Mahalle is PhD from Aalborg university and is IEEE member, ACM member, Life member ISTE and graduated in Computer Engineering from Amravati University, Maharashtra, India in 2000 and received Master in Computer Engineering from Pune University in 2007. From 2000 to 2005, was working as Assistant Professor in Vishwakarma Institute of technology, Pune, India. From August 2005, he is working as Professor and Head in Department of Computer Engineering, STES's Smt. Kashibai Navale College of Engineering, Pune, India. He published 42 research publications at national and international journals and conferences. He has authored 5 books on subjects like Data Structures, Theory of Computations and Programming Languages. He is also the recipient of "Best Faculty Award" by STES and Cognizant Technologies Solutions. He has guided more than 100 plus undergraduate students and 10 plus post-graduate students for projects. His research interests are Algorithms, IoT, Identity Management and Security. He has also delivered invited talk on "Identity Management in IoT" to Symantec Research Lab, Mountain View, California.

Poonam N. Railkar received the Master in Computer Engineering (Computer Networks) from Pune University Maharashtra, India in the year 2013. From September 2012, she is currently working as an Assistant Professor in Department of Computer Engineering, STES's Smt. Kashibai Navale College of Engineering, Pune, India. She has published 10 plus papers at national and international journals and conferences. She has guided more than 5 plus under-graduate students for projects. Her research interests are Mobile Computing, Identity Management, Security and Database Management System Applications.